SpringerBriefs in Earth System Sciences

Series Editors

Gerrit Lohmann, Universität Bremen, Bremen, Germany

Lawrence A. Mysak, Department of Atmospheric and Oceanic Science, McGill University, Montreal, QC, Canada

Justus Notholt, Institute of Environmental Physics, University of Bremen, Bremen, Germany

Jorge Rabassa, Labaratorio de Geomorfología y Cuaternar, CADIC-CONICET, Ushuaia, Tierra del Fuego, Argentina

Vikram Unnithan, Department of Earth and Space Sciences, Jacobs University Bremen, Bremen, Germany

SpringerBriefs in Earth System Sciences present concise summaries of cutting-edge research and practical applications. The series focuses on interdisciplinary research linking the lithosphere, atmosphere, biosphere, cryosphere, and hydrosphere building the system earth. It publishes peer-reviewed monographs under the editorial supervision of an international advisory board with the aim to publish 8 to 12 weeks after acceptance. Featuring compact volumes of 50 to 125 pages (approx. 20,000—70,000 words), the series covers a range of content from professional to academic such as:

- A timely reports of state-of-the art analytical techniques
- bridges between new research results
- snapshots of hot and/or emerging topics
- literature reviews
- in-depth case studies

Briefs are published as part of Springer's eBook collection, with millions of users worldwide. In addition, Briefs are available for individual print and electronic purchase. Briefs are characterized by fast, global electronic dissemination, standard publishing contracts, easy-to-use manuscript preparation and formatting guidelines, and expedited production schedules.

Both solicited and unsolicited manuscripts are considered for publication in this series.

More information about this series at https://link.springer.com/bookseries/10032

Manuel Enrique Pardo Echarte ·
Osvaldo Rodríguez Morán ·
Lourdes Jiménez de la Fuente ·
Jessica Morales González · Orelvis Delgado López

Geological-Structural Mapping and Favorable Sectors for Oil and Gas in Cuba

Non-seismic Exploration Methods

 Springer

Manuel Enrique Pardo Echarte
Geology Basic Science and Technology
Unit, ESRU
Petroleum Research Center
El Cerro, La Habana, Cuba

Lourdes Jiménez de la Fuente
Geology Basic Science and Technology
Unit, ESRU
Petroleum Research Center
El Cerro, La Habana, Cuba

Orelvis Delgado López
Geology Basic Science and Technology
Unit, ESRU
Petroleum Research Center
El Cerro, La Habana, Cuba

Osvaldo Rodríguez Morán
Geosciences Department
Technological University of La Habana
Marianao, La Habana, Cuba

Jessica Morales González
Geosciences Department
Technological University of La Habana
Marianao, La Habana, Cuba

ISSN 2191-589X ISSN 2191-5903 (electronic)
SpringerBriefs in Earth System Sciences
ISBN 978-3-030-92974-9 ISBN 978-3-030-92975-6 (eBook)
https://doi.org/10.1007/978-3-030-92975-6

This Springer imprint is published by the registered company Springer Nature Switzerland AG
The registered company address is: Gewerbestrasse 11, 6330 Cham, Switzerland

Foreword

After a little more than the first billion years, from the first stage of consolidation of our planet, the movement of the tectonic plates began to govern its evolution. This mechanism of global planetary dynamics would lead to the intensification of the recycling of terrestrial materials, climate changes and biological diversification on Earth. Thus, a particular geodiversity and biodiversity appear as the distinctive signature of our current world.

In the region that we know as the Caribbean, the processes and mechanisms of physical and biological evolution have left their mark since at least the beginning of the Mesozoic and up to the present day. With the birth of the "Tethys Sea", between the continental, North American and South American margins, the first magmatic and sedimentary rocks of the primitive Caribbean emerged. The Proto-Caribbean region would occupy an area, perhaps a million square kilometers in its beginnings; to develop and reduce itself to the present day, perhaps to a space barely equivalent to a quarter or a fifth. Subduction would lead to the disappearance of most of the materials that existed here; the tectonic accretion of the lands belonging to some paleogeographic domains against others, would lead to the shortening of the regional profile, and at the same time, to its thickening in parts.

The Cuban geological substrate, originated during the evolution of this intense geodynamic context, represents a complex, highly diverse set of lithic materials reduced to a geotectonic and structural strip, attached to the southern margin of the North American continent. Here the remains of several paleogeographic domains converged. Namely, elements of the intraoceanic volcanic island arc systems, which were born, developed and disappeared during the Cretaceous and Paleogene periods; relics of oceanic crusts; of intraoceanic rift systems and associated sediments; of metaphorized continental lands; intramontane and piggy backs basins; of foreland basins; and postorogenic coverage until recent. As is understandable, the original relationships between the aforementioned geological–geographical elements are today very difficult to restore in a coherent reconstruction. Thus, Cuban geosciences have had, historically, and to date, immense cognitive challenges: establishing regularities and useful hypotheses in economic practice, such as the exploration of natural resources essential for the development of our country.

Today the geological substrate of the Cuban territory is assumed, in general, as parallel tectonostratigraphic strips, more or less narrow, ridged and folded, imbricated in several mantles, forming complex structural stacks, in a sublatitudinal direction, segmented and accreted along from the southern margin of the North American continent—Bahamas platform to the north, and from the Mayan continental block—Yucatan platform to the west. Under these conditions, hydrocarbon exploration in particular faces the compression of past oil systems, their evolutions and, most importantly, the establishment of their elements in the current geotectonic situation and structure. The identification of source rocks, the types of original organic matter, the conditions of genesis for the oils and their migrations and dismigrations over time; the types of traps and the nature of the reservoirs, the flows and migration routes; and local and regional seals, the possible volumes of hydrocarbons contained, among others, etc., are goals assumed by generations of Cuban and foreign geoscientists, already for almost a century of work.

The surface shows, tar springs and natural asphalt, were the direct indicators of the existence of hydrocarbons in a given area, but the search for economically significant accumulations, through the drilling of exploratory wells, in occluded deposits, would be the most promising challenge. But only the knowledge of the geological premises was quickly insufficient to obtain optimal results in oil exploration. Thus, geophysical methods decisively complete the answer to such challenges. The diversification of its applications, based on natural physical fields, had promising advances for exploration. Well logging methods reinforced the interpretation of lithologies, their properties, limits and thicknesses of different units traversed, allowing the characterization of reservoirs and testing the most promising correlations.

At present, the theoretical and technological advances that are presented to us are truly impressive, however, the increasing depth of exploration, the already described high geological complexity of the Cuban territory, the surprising variability in the types of reservoirs (carbonates, serpentinites, volcanic rocks and others) and the need to achieve high economic efficiency, among other pressures, require a complex, holistic approach to current gas-oil exploration work. Thus, the integration, increasingly sophisticated and creative with the use of new software, of all available data and methods, is shown as the most advisable way to search for oil and gas in Cuba. The research team that offers us this scientific contribution dedicates their efforts to this cardinal objective.

The authors value that the investigations in oil and gas exploration take into account the most classical components that characterize the geological-geophysical-geochemical "macro" aspects, those that are more evident and better accepted. However, they maintain that there are indicators considered "micro", less evident and not widely accepted, but that they can play an important complementary role to the first, with the common and main objective of achieving greater efficiency in new discoveries. To solve this last task, they assume the so-called unconventional geophysical-geochemical exploration methods. The authors base their line of study, developed in decades of work, on the existence of microseepage of light hydrocarbons in the vertical, which lead to superficial modifications on accumulations in the subsoil, such as seismic structures or stratigraphic traps, charged. Hence, the need to achieve a detection and mapping of the areas affected by microseepage; and, in addition, contribute to recognize the geological–structural context of its occurrences.

It is known how expensive it is to acquire and process high-quality modern seismic information, carried out with the use of sophisticated and highly specialized technology. On the other hand, the present team of researchers, in order to fulfill their task, promote the use of non-seismic exploration methods used in Cuba: remote sensing, gravimetry, aeromagnetometry, airborne gamma spectrometry (AGS) and morphometry (non-conventional) from Digital Elevation Model 90 × 90 m. They clarify that the AGS also classifies as an unconventional geophysical-geochemical method along with the Redox Complex. In addition, they process gravimetry, aeromagnetometry and morphometry at a scale of 1:50000, AGS at a scale of 1:100000, and ASTER and LANDSAT satellite images (with resolution of 15 and 30 m, respectively). The result to be achieved is the reduction of areas for more effective exploration, which also leads to risk reduction. From the implementation of these methods, the perspective sectors for oil and gas are obtained once the integration with the corresponding geology and seismic has been carried out.

In the western region of Cuba, with several land exploration blocks, geological-structural cartography, based on gravi-magnetic and morphometric data, establishes possible structural highs that alternate with depressed areas. The authors achieve a version of the mapping of favorable sectors of interest related to conventional oil and gas in the Tectonostratigraphic Units of Camajuaní and Placetas, associated with the Continental Margin of North America. The proposal is based on the presence of a complex of anomalies of geophysical-morphometric indicators, considering minimums of the K/Th ratio and local maximums of U (Ra) in its periphery, remote sensing anomalies, as well as local gravimetric, magnetic and morphometric maxima.

In areas with microseepage near the Motembo oil field and the Menéndez-Bolaños area, in the province of Villa Clara, they were corroborated by data from surface geochemical surveys with the presence of gas in soils. These areas of possible hydrocarbon microseepage were achieved from the study of the latest indirect indices through the analysis of optical images. In another case, in the Majaguillar area, another area with microseepage was identified, with similar patterns, supported by the interpretation of a complex of non-seismic and non-conventional methods.

The well-known Central Basin of Cuba has been one of the important oil regions of the country, distinguished by its peculiarities in the exploration and extraction of hydrocarbons, given by reservoirs in effusive and clastic volcanic rocks. This is a first-order structure, regional in nature, it had its development in relation to the evolution of the La Trocha fault system. This last tectonic element has had a long existence, dividing the substratum of the Cretaceous volcanic arc, participating in regional thrust and folding and influencing the diastrophism of the crust, elevation and subsidence of blocks, erosion and sedimentary accumulation, regional, etc. In this context, the authors point out that the 2D physical-geological models of potential fields have the objective of clarifying the deep constitution of the territory, given the interest in finding conventional oil and gas from the Tectonostratigraphic Units of Camajuaní and Placetas, in the sediments of the Continental Margin of North America. From this it is inferred that the hydrocarbons known in the area are products of migration from the underlying bedrock, related to the basin-platform slope sequences. Several sectors are defined as perspective, to be contrasted with the seismic information.

The authors evaluated several areas that meet, according to their results, the geological-oil premises for the presence of large accumulations of high-quality oil. For their study, they used more than 2000 rock samples, more than 200 oil for physicochemical analysis and biomarker testing, and 27 organic extracts of bedrock for geochemical correlations. Only those areas with high thermal evolution of oil were taken. As a result, they determined the types of oils of families I and II of Cuban crude oils, which indicate the presence in the subsoil of an Upper Jurassic-Lower Cretaceous tectonic plate; and Family III, indicating the presence in the subsoil of the Lower-Upper Cretaceous tectonic plate. Among the areas analyzed, there are exploratory blocks in the Central Basin, which meets all the premises for the presence of large accumulations of high-quality oil. According to the work team, in the rest of the assessed areas there is variation in thermal maturity, increasing the risk of finding deep accumulations with high-quality commercial crude.

For more than three decades, the main authors (MEPE, ORM) have accumulated vast experience in the application of the most varied geophysical methods, in geophysical-geochemical mapping, in geological-structural and tectonic regionalization, among others. The development of techniques and methods for studying the Redox Potential with different uses; and in the last two decades, the integration of the Redox Complex with its introduction in research associated with the exploration of hydrocarbons in the country and other regions. The foregoing has led to a solid conceptual foundation and innovation in integrative methods and techniques, with holistic interpretations and demonstrated practical scope. They even have experience in the development of their own software, to solve tasks in their daily work.

The present volume *Geological-Structural Mapping and Favourable Sectors for Oil and Gas in Cuba. Non-seismic Exploration Methods* groups and consolidates a new vision on the knowledge in the integration of data and information, applicable in the exploration of hydrocarbons in Cuba. The extensive experience of the main authors of the work team, both as researchers in different areas of application of geophysical-geochemical-geological studies, as well as professors in higher education centers in Cuba and abroad, they endorse the consistent scientific conceptualization of the content that is offered to us. The young researchers, accompanying those wise men, broaden the horizons of the approaches used in the studies carried out and provide comprehensiveness to the consolidated results.

Finally, I would like to congratulate the guides and participants in this book, and a call to attention to those decision makers, responsible for supporting the application of the methods, techniques and concepts demonstrated here reliably. I thank Dr. Manuel Pardo Echarte for the privilege of having the opportunity to review this valuable contribution.

El Cerro, La Habana, Cuba

Dr. Reinaldo Rojas Consuegra
Head of Regional Geology Department,
Geology Scientific-Research Unit,
Centro de Investigación del Petróleo

Preface

In solving any problem of geological exploration, it is essential to use a holistic approach, that is, to consider the integration of the parts or components subject to investigation. As a rule, these components characterize the "macro" geological-geophysical-geochemical aspects (large, more evident and better accepted) and the "micro" ones (small, less evident and little accepted). Usually, for the study of the latter, the so-called non-conventional geophysical-geochemical exploration methods are used. In the particular case of oil and gas exploration, the nucleus of the "micro" aspects is characterized by the microseepage of light hydrocarbons, with a vertical nature on the gas-oil accumulations, as well as, by the modifications that occur in the superficial medium as a result of it. Thus, the problem lies in the need of detection and mapping of microseepage, complementing the conventional methods with a valuable information: the possible hydrocarbon charge in a seismic structure and/or the presence of subtle stratigraphic traps. Also, it is of interest to know the geological-structural framework where microseepage occur. The non-seismic exploration methods used in Cuba are: remote sensing, gravimetry, aeromagnetometry, airborne gamma spectrometry (AGS) and morphometry (non-conventional, from the Digital Elevation Model 90 × 90 m). The AGS also classifies, as a non-conventional geophysical-geochemical method, together with the Redox Complex. In the work, gravimetry, aeromagnetometry and morphometry at a scale of 1:50000, AGS at a scale of 1:100000, and ASTER and LANDSAT satellite images (with 15 and 30 m resolution, respectively) are processed. The purposes of the referred complex of methods are to reduce areas, increasing the effectiveness of exploration with a decrease in its risks. From the implementation of these methods, perspective sectors for oil and gas are obtained once the integration with geology and seismic has been carried out. The work presents a brief theoretical account of the methods and, as practical results, a set of perspective sectors of possible interest for exploration.

It is known that non-seismic exploration methods offer necessary and important information on the geological-structural mapping of the territories and on the presence in them of vertical areas of active microseepage of light hydrocarbons, witnesses to possible accumulations at depth. That is why the benefits of using these methods, prior to their integration with geological and seismic data, translate into a first approximation, valid for an initial understanding of geology and mapping of favorable areas of possible gas-oil interest. Such are the objectives of the investigation at the western Cuba region (land exploration blocks 6, 7, 8A and 9A). To meet these objectives, gravimetry and aeromagnetometry at a scale of 1:50000 and 1:250000, AGS at a scale of 1:100000, (ASTER satellite images) and the Digital Elevation Model 90 × 90 m of the territory were processed. For the geological interpretation, the Digital Geological Map of the Republic of Cuba at a scale of 1:100000 was used. The geological-structural cartography of the study region, based on the gravi-magnetic and morphometric data, establishes possible structural highs which alternate with depressed areas. A version of favorable sectors mapping of gas-oil interest (linked to conventional oil and gas from the Camajuaní and Placetas Tectonostratigraphic Units) is based on the presence of a complex of geophysical-morphometric indicator anomalies. It considers: minimums of the K/Th ratio and the local maxima of U (Ra) in its periphery; remote sensing anomalies as well as; local gravimetric, magnetic and morphometric maxima. The work presents an account of the processing and interpretation of non-seismic exploration methods and, as practical results, the foundation of the main favorable sectors of possible interest for exploration. The geological-structural cartography of the study region, based on gravi-magnetic and morphometric data, allowed to clarify the structural picture of the territory, where geological structures of Cuban direction (SE-NW course) and others of SW-NE and latitudinal (EW) course predominate.

The remote sensors allow the analysis of the terrain in order to identify indications of possible hydrocarbon microseepage in soils and sediments. Microseepage are invisibly hydrocarbon escapes that are manifested on the surface through changes in the reflectance, stress on vegetation, abnormal concentrations of kaolinite, iron oxides and carbonate alterations. The objective of the work is to identify areas of possible hydrocarbon microseepage from the study of the last four indirect indices through the analysis of optical images. The digital processing of multispectral images, Aster, Landsat 7 and 8 consisted of Red Green Blue (RGB) combinations, band ratios and integration and analysis of the information in the Geographic Information System. Spatial-temporal studies of Normalized Difference Vegetation Index (NDVI), analysis of thermal reflectivity images of the surface related to the stress of the vegetation and studies of mineralogical anomalies were carried out at a scale of 1:50000. Areas with microseepage near the Motembo oilfield and the zone of Menéndez in Villa Clara province were interpreted, being corroborated by data from surface geochemical surveys with the presence of gas in soils. Another area with microseepage was identified in the Majaguillar area, with similar patterns to the previous ones, which was supported by the interpretation of a complex of non-seismic and non-conventional methods. In addition to these, other areas were interpreted with a lower degree of confidence but with characteristics very similar to those already established.

The Central Basin of Cuba was the largest oil-producing region in the country during the 1960s. However, after the 1990s with the discovery of the Pina oilfield, there has been no other significant discovery. Exploration failures are considered to be conditioned, in part, by the high geological complexity of the region and by the volcanic nature of the sequences present, which limit the depth of reflection seismic research. Thus, the problem lies in the need to use 2D physical-geological modeling of potential fields in order to help clarify the deep constitution of the territory, given the interest for finding conventional oil and gas from the Tectonostratigraphic Units (TSUs) of Camajuaní and Placetas at the sediments of the North American Continental Margin. Geological and petrophysical data, seismic data and potential fields of the northeastern region of the Central Basin of Cuba were evaluated in the preparation and interpretation of three 2D models of potential fields: one that is longitudinal to the basin and two that cut the Cristales and Pina oilfields, respectively. As a result, from the 2D physical-geological models, the hypothesis of the existence in the whole basin of carbonates from the North American Continental Margin, considered as source rock, is validated. According to the models, the top of these rocks is located, at the Pina sector, between 2.98–4.3 km, while at the Jatibonico-Cristales and Catalina sectors they range between 5.55–6.6 km and 6.2 km, respectively. In addition, their thickness decreases from north (5 km) to south (1.3 km) and, conversely, the one of Zaza Terrain. This reinforces the hypothesis of the best prospects for finding conventional oil from the Camajuaní and Placetas TSUs in the Pina sector.

Cuba has been producing oil since 1936. The first fields produced good-quality crude, but with little production due to bad reservoirs (ophiolites). Later, better reservoirs were discovered (Mesozoic carbonates) with large resources, but with poor quality oil. So far, no large reserves of high-quality oils have been discovered in Cuba. The objective of this study was: to evaluate areas that meet the petroleum-geologist premises for the presence of large accumulations of high-quality oil. For that reason, 2038 rock samples (for rock-eval studies), 207 oil samples (for physicochemical and biomarkers analysis) and 27 organic extracts from source rock (for oils— source rock correlations) were used. The petroleum systems exploratory method was followed. Only those areas with high thermal evolution oils were taken into account (Exploratory blocks 21-23, Block 7, Block 6, Block 13 and Blocks 17-18). It is concluded that the presence of families I and II of Cuban oils indicates the presence of the J_3-K_1 tectonic sheet; Family III of Cuban oils, indicates the presence of the K_1-K_2 tectonic sheet. The premises that an area must meet for the existence of large accumulations of high-quality oil are: oil with high thermal evolution, rich in sulfur and protected from biodegradation, Veloz Group-type reservoirs. Of the areas evaluated, Block 21-23 is the one that meets all the premises for the presence of large accumulations of high-quality oil. In the rest, there is variation in thermal maturity, increasing the risk of finding deep accumulations with high-quality commercial crude, and also, presence of reservoirs that have not demonstrated large reserves (Camajuaní Tectonostratigraphic Unit).

El Cerro, Cuba   Manuel Enrique Pardo Echarte
Marianao, Cuba   Osvaldo Rodríguez Morán
El Cerro, Cuba   Lourdes Jiménez de la Fuente
Marianao, Cuba   Jessica Morales González
El Cerro, Cuba   Orelvis Delgado López

Contents

Abbreviations

AAC	Ascending analytical continuation
AC	Anomalous complex
AGS	Airborne gamma spectrometry
°API	American Petroleum Institute, oil density measurements unit
CEINPET	Centro de Investigación del Petróleo
CNOB	Cuban North Oil Belt
CNTFB	Cuban North Thrusted and Folded Belt
CUPET	Unión Cuba Petróleo
CVA	Cretaceous Volcanic Arcs
DEM	Digital Elevation Model
DEMreg500	Digital Elevation Model, regional component from AAC at 500 m
DEMres500	Digital Elevation Model, residual component from AAC at 500 m
DEMTHD	Digital Elevation Model, total horizontal derivative
DT	Aeromagnetometry and/or magnetic field
DT250rp	Reduced to the pole magnetic field, 1:250000 scale
DT250rpTHD	Total horizontal derivative of reduced to the pole magnetic field, 1:250000 scale
DT250rpVD	First-order vertical derivative of reduced to the pole magnetic field, 1:250000 scale
DTrp (RP)	Reduced to the pole magnetic field, 1:50000 scale
DTrpreg500	Reduced to the pole magnetic field, regional component from AAC at 500 m, 1:50000 scale
DTrpres12000	Reduced to the pole magnetic field, residual component from AAC at 12000 m
DTrpTHD	Total horizontal derivative of reduced to the pole magnetic field, 1:50000 scale
DTrpVD	First-order vertical derivative of reduced to the pole magnetic field, 1:50000 scale
ETM	Enhanced Thematic Mapper
FLAASH	Fast line of sight atmospheric analysis of hypercubes
Fm	Geological formation

Gb	Gravimetry and/or Bouguer gravity field
Gbres12000	Bouguer gravity field, residual component from AAC at 12000 m
Gbres500	Bouguer gravity field, residual component from AAC at 500 m
GbTHD	Total horizontal derivative of Bouguer gravity field
GbVD	First-order vertical derivative of Bouguer gravity field
GIS	Geographic Information System
GPC	Gabbro-Plagiogranite Complex
IDP	Image digital processing
It	Total gamma intensity
LST	Land surface temperature
MMI	Metal mobile ion
NACM	North American Continental Margin
NCGGEM	Non-conventional geophysical-geochemical exploration methods
NDVI	Normalized Difference Vegetation Index
NIR	Near infrared
NSNCEM	Non-seismic and non-conventional exploration methods
OA	Ophiolite Association
OC1	Oeste de Ceballos 1
OC2	Oeste de Ceballos 2
OLI/TIRS	Operational Land Imager/ thermal infrared sensor
PVT	Pressure volume temperature
Red	Visible red
RGB	Red–green–blue
RGM	Response Generalized Model
RS	Remote sensing
RSR	Reduced spectral reflectance
SWIR	Short wave infrared
TDR	Magnetic field inclination derivative
TIR	Thermic infrared
TSU	Tectonostratigraphic Unit
VIS	Visible region of the electromagnetic spectrum

List of Figures

List of Tables

Chapter 1
Non-seismic and Non-conventional Exploration Methods for Oil and Gas in Cuba: Perspective Sectors

Manuel Enrique Pardo Echarte, Osvaldo Rodríguez Morán, Jessica Morales González, and Lourdes Jiménez de la Fuente

Abstract In solving any problem of geological exploration, it is essential to use a holistic approach, that is, to consider the integration of the parts or components subject to investigation. As a rule, these components characterize the "macro" geological–geophysical–geochemical aspects (large, more evident and better accepted) and the "micro" ones (small, less evident, and little accepted). Usually, for the study of the latter, the so-called non-conventional geophysical–geochemical exploration methods are used. In the particular case of oil and gas exploration, the nucleus of the "micro" aspects is characterized by the microseepage of light hydrocarbons, with a vertical nature on the gas–oil accumulations, as well as, by the modifications that occur in the superficial medium as a result of it. Thus, the problem lies in the need of detection and mapping of microseepage, complementing the conventional methods with a valuable information: the possible hydrocarbon charge in a seismic structure and/or the presence of subtle stratigraphic traps. Also, it is of interest to know the geological–structural framework where microseepage occurs. The non-seismic exploration methods used in Cuba are: remote sensing, gravimetry, aeromagnetometry, airborne gamma spectrometry (AGS) and morphometry (non-conventional, from the Digital Elevation Model 90 × 90 m). The AGS also classifies, as a non-conventional geophysical–geochemical method, together with the Redox Complex. In the work, gravimetry, aeromagnetometry, and morphometry at a scale of 1:50,000, AGS at a scale of 1:100,000, and ASTER and LANDSAT satellite images (with 15 and 30 m resolution, respectively) are processed. The purposes of the referred complex of methods are to reduce areas, increasing the effectiveness of exploration with a decrease in its risks. From the implementation of these methods,

M. E. Pardo Echarte (✉) · L. Jiménez de la Fuente
Centro de Investigación del Petróleo, Churruca, No. 481, Vía Blanca y Washington, 12000 El Cerro, La Habana, CP, Cuba
e-mail: pardo@ceinpet.cupet.cu

L. Jiménez de la Fuente
e-mail: lourdes@ceinpet.cupet.cu

O. Rodríguez Morán · J. Morales González
Universidad Tecnológica de La Habana "José Antonio Echeverría" (Cujae) Calle, 114 no. 11901 entre Ciclo Vía y Rotonda, 11500 Marianao, La Habana, CP, Cuba

perspective sectors for oil and gas are obtained once the integration with geology and seismic has been carried out. The work presents a brief theoretical account of the methods and, as practical results, a set of perspective sectors of possible interest for exploration.

Keywords Non-seismic and non-conventional methods of oil and gas exploration · Remote sensing · Gravimetry · Aeromagnetometry · Airborne gamma spectrometry · Morphometry · Digital elevation model · Redox complex · Cuba

Abbreviations

AGS	Airborne gamma spectrometry
DEM	Digital elevation model
NSNCEM	Non-seismic and non-conventional exploration methods
NCGGEM	Non-conventional geophysical–geochemical exploration methods
AC	Anomalous complex
It	Total gamma intensity
Gb	Gravimetry
DT	Aeromagnetometry
RS	Remote sensing
RSR	Reduced spectral reflectance
AAC	Ascending analytical continuation
GbVD	First vertical derivative of gravity field
GbTHD	Total horizontal derivative of gravity field
TSU	Tectonostratigraphic unit
RGM	Response generalized model
DTrp (RP)	Reduced to the pole magnetic field
DTrpVD	First vertical derivative of reduced to the pole magnetic field
DTrpTHD	Total horizontal derivative of reduced to the pole magnetic field
TDR	Magnetic field inclination derivative
MMI	Metal mobile ion
OC2	Oeste de Ceballos 2
OC1	Oeste de Ceballos 1

1.1 Introduction

It is well documented that most hydrocarbon accumulations have leaks or microseepage, which are predominantly vertical, as well as that they can be detected and mapped using various non-seismic and non-conventional exploration methods (NSNCEM). Given the direct link with the mapping of possible hydrocarbon microseepage, the reduction of areas in the application of these methods, for the purposes of seismic planning, is much more effective. Schumacher (2014) has stated that data on hydrocarbon microseepage—when properly acquired, interpreted, and integrated with geological, seismic and other conventional methods data—lead to a better evaluation of prospective areas and exploration risks.

In solving any problem in geological exploration, it is essential to use a holistic approach during the investigation, that is, the integration of its investigative-methodological parts or components. These components characterize, in turn, the "macro" geological–geophysical–geochemical aspects (large, more evident and better accepted) and the "micro" (small, less evident and little accepted). For the study of the latter, the so-called non-conventional geophysical–geochemical exploration methods (NCGGEM) are used. In the particular case of oil and gas exploration, the center of the "micro" aspects is constituted by the microseepage of light hydrocarbons, with a vertical character, on the gas–oil accumulations and the modifications that occur as a result of it in the surface environment, fundamentally in soils. Among these modifications are: increase in the content of metallic elements (V, Ni, Fe, Pb and Zn); increase in Magnetic Susceptibility; decrease in Redox Potential and reduced Spectral Reflectance; presence of minima of the K/Th ratio with maximums of U(Ra), in its periphery in a majority way; presence of subtle geomorphic residual maximums and remote sensing anomalies, among others. The NCGGEM are in charge of detecting and mapping these subtle modifications in the surface environment, thus complementing the complex of conventional methods with valuable information: the possible hydrocarbon charge in a seismic structure and/or, the presence of subtle stratigraphic traps. Unfortunately, the acceptance of the NCGGEM and its results in the exploration of oil and gas are still going through a bad time, since the community of explorers and geoscientists who use seismic as their main investigative-exploratory tool, still reject them.

The non-seismic exploration methods used in Cuba, considered in this work, are: sensing, gravimetry, aeromagnetometry, airborne gamma spectrometry and morphometry. The penultimate, it also classifies, as a non-conventional geophysical–geochemical method, together with the Redox Complex. The aforementioned complex has as a novelty the integration of methods, all based, with the exception of gravimetry, on the anomalous physical–chemical response of the environment in front of the hydrocarbon microseepage on the accumulations in depth, since its purpose is to reduce areas and the elevation of the effectiveness of the exploration, with a more sensible reduction of its risks.

The non-seismic exploration methods lead, during the reduction of areas, to the mapping of not a few favorable sectors, given by groups of possible hydrocarbon microseepage, which have to be recognized and evaluated on land later, by the Redox Complex. Only very few perspective sectors are derived from this, where the possible presence of hydrocarbons in depth is established, fundamentally from the anomalous indications of the Soil-geochemistry (chemical elements of vanadium and nickel).

Thus, this work aims, based on a rigorous and exhaustive systematization and generalization of all the NSNCEM's information used in Cuba, to present a brief theoretical account of these methods and as practical results, the main perspective sectors of possible interest for exploration. Most of these sectors have seismic and/or well information to support them.

As antecedents, in the period 2014–2018, NSNCEM's GeoSoft Oasis Montaj Projects corresponding to onshore oil blocks (6, 7, 8A, 9, 9A, 13, 14, 17, 18, 21, 21A and 23) many of which were later revised, completed and detailed. Some of their results were presented in Pardo Echarte and Cobiella Reguera (2017) and Pardo Echarte et al. (2019).

The article presents: an Introduction with a heading of Physical–Chemical–Geological Premises; Materials and Methods with an introduction to the NSNCEM used in Cuba; Results and Discussion, with the perspective sectors for three separate regions: Occidental (surroundings of the Motembo oilfield), Central (surroundings of the Catalina-Cristales and Pina-Ceballos regions, in the Central Basin) and Central-Eastern (surroundings of Maniabón, region of Puerto Padre). The description includes the presentation of the anomalous complex (AC), for the non-seismic methods and the profile of the Redox Complex (where available), with its comments. The seismic information is not presented here, given its confidential nature.

1.1.1 Physical–Chemical–Geological Premises

From the point of view of surface geochemistry, according to Price (1985), Schumacher (1996) Saunders et al. (1999) and Pardo and Rodríguez (2016), the physical–chemical–geological premises that support the application of non-conventional geophysical–geochemical–morphometric exploration methods are the following:

As the light hydrocarbons rise from the accumulation, bacterial oxidation produces, as a by-product, carbonic and organic acid, as well as hydrogen sulfide. For its part, carbonic acid reacts with clay minerals, destroying them, while creating secondary carbonate mineralization and silicification. Close to the surface, both materials are denser and more resistant to erosion, with an effect in the increase of the seismic velocity on the accumulation, as well as in the formation of erosional topographic maxima.

Regarding oil and gas accumulations, the decomposition of clays in the soils as a result of the microseepage of light hydrocarbons is responsible for the minimum radiation observed: Potassium is leached from the system toward the edges of the vertical projection of the accumulation, where it precipitates resulting in a "halo" of high values. Thorium remains relatively fixed, in its original distribution within insoluble heavy minerals, hence minimums of the K/Th ratio surrounded by maximums are observed on these deposits. In the majority of the periphery of these anomalies, local increases in U(Ra) are also observed. Additionally, the aforementioned relationship offers the opportunity to eliminate a series of undesirable effects on spectrometric measurements (influence of lithology, humidity, vegetation and measurement geometry).

Regarding the role of hydrogen sulfide, its own presence conditions the formation of a reducing environment column (minimum of the Redox Potential) on the accumulation. This reducing environment favors, in turn, the conversion of non-magnetic iron minerals into more stable magnetic (diagenetic) varieties such as magnetite, maghemite, pyrrhotin and greigite, all responsible for the increase (maxima) of susceptibility. Magnetics of rocks and soils on accumulation; a fact that explains the observed inverse correlation between both attributes (minima of the Redox Potential and maxima of the Magnetic Susceptibility) and justifies the integration of the methods. The arrival to the surface of the metallic ions contained in the hydrocarbons (V, Ni, Fe, Pb and Zn, among others) conditions the presence of a subtle anomaly of these elements in the soil and a slight change in its coloration (darkening), which is reflected by anomalies of the Reduced Spectral Reflectance, evidences that justify, also, the integration of these techniques (Pardo Echarte and Rodríguez Morán 2016).

1.2 Materials and Methods

1.2.1 Materials

The materials used and their sources are the following:

- Gravimetric and magnetic field grids at 1:50,000 scale and airborne gamma spectrometry (channels: It, U, Th and K) at 1:100,000 scale of the Republic of Cuba (Mondelo Diez et al. 2011).
- Digital Elevation Model (DEM) (90 × 90 m) taken from Sánchez Cruz et al. (2015), with source at: http://www.cgiar-csi.org/data/srtm-90m-digital-elevation.
- ASTER and LANDSAT 7 satellite images of the Republic of Cuba and their interpretation (Jiménez de La Fuente and Pardo Echarte 2020).
- Digital maps of the hydrocarbons shows and the oil wells of the Republic of Cuba at a scale of 1:250,000 (Colectivo de Autores 2008, 2009 respectively).
- Digital Geological Map of the Republic of Cuba at a scale of 1:100,000 (Colectivo de Autores 2010), used for the purposes of geological interpretation.

The geophysical information processing was carried out using the GeoSoft Oasis Montaj version 7.01 software. The processing and interpretation of the Redox Complex data were carried out with the redox system (database and software), developed in 2005 (Rodríguez Morán 2005).

1.2.2 Methods

The non-seismic geophysical exploration methods used in Cuba are:

- Remote Sensing (RS)
- Gravimetry (Gb)
- Aeromagnetometry (DT)
- Airborne gamma spectrometry (AGS). This is also classified as a non-conventional geophysical–geochemical method together with the Redox Complex.

1.2.2.1 Remote Sensing

In an area where there are hydrocarbon microseepage, an acid-reducing medium takes place, which allows the layers with clays and iron to be altered. Thus, for example, clays are altered from montmorillonitic to kaolinitic. For its part, ferric iron (Fe^{+++}) becomes ferrous iron (Fe^{++}). A secondary carbonation process also occurs. All these alterations of the environment are reflected by tonal anomalies in the satellite images.

The non-conventional way of processing the ASTER (15 m resolution) and LANDSAT 7 (30 m resolution) images, indistinctly, is related to the subtle darkening of the ground due to the presence of metals, which is expressed in weak negative residual spectral reflectance anomalies within the Visible Band (band 1). These anomalies are reflected in the so-called redox scenarios. In the processing, band 1 is converted to Spectral Reflectance values, when analyzing the background values (averages) for the area where the target is located and this value is subtracted from the entire area. As a result, an image with residual reflectance values (transformed) is obtained, which is conveniently reclassified to highlight the anomalous area (negative values). The referred procedure constitutes an innovation generalization of the results of the Redox Complex regarding the use of Reduced Spectral Reflectance (RSR) in soil samples.

1.2.2.2 Gravimetry

The application of the gravimetric method offers the possibility of studying the regional geological constitution, with better results for folded belts, such as Cuba, by allowing, in this way, tectonic regionalization, geological–structural mapping of large units and the location of structures in the sedimentary cover. It is, then, an effective means of mapping sedimentary basins and the main tectonic features with which various mineral and energy resources are sometimes linked. From the more local point of view, it is accepted for the location and mapping of salt, reef, granite and ultrabasic bodies (Pardo Echarte and Cobiella Reguera 2017).

The gravitational field (Bouguer Reduction, 2.3 t/m^3) is subjected to regional–residual separation from the Ascending Analytical Continuation (AAC) for the heights of 500, 2000 and 6000 m, given by the order of depth of the possible oil and gas targets and seismic studies. For geological–structural mapping, the first vertical derivative (GbVD) and the total horizontal derivative (GbTHD) are used, as the latter allows the mapping of the different tectonic alignments (faults) present (Pardo Echarte et al. 2019).

The geological–structural mapping based on the regionalization of the GbVD field distinguishes different zones of intense maximum values, maximums, intermediate values, minimum and intense minimums. Intense maxima and maxima can, as a rule, be associated with the presence of carbonates and ophiolites, respectively. The intense minima and maxima can be associated with the sediments of the basins and the sequences of the Camajuaní and Placetas Tectonostratigraphic Units (TSUs). The salt structures mapped by the AGS and the DEM are revealed by minimum values of the GbVD field.

The residual field at 500 m (Gbres500) and the GbVD allow the mapping of very subtle local gravimetric maxima associated with structural uplifts of carbonates and/or volcanic, with gas–oil interest (e.g., Pina and Cristales oilfields).

1.2.2.3 Aeromagnetometry

This method offers an aid to geological cartography of volcano-sedimentary and intrusive formations in volcanic arches, such as those in the territory of Cuba. In the presence of non-magnetic sedimentary rocks, the aeromagnetic survey data provide information on the nature and depth of the basic–ultrabasic and/or crystalline basement. Locally, the ability to map geological–structural features is reinforced by the possibility of detecting anomalies of very low amplitude associated with potential hydrocarbon microseepage, in addition to the fact that intrusive (granitoid) and protusive (ophiolite) bodies are often distinguished from direct form (Pardo Echarte and Cobiella Reguera 2017).

The magnetic field undergoes pole reduction (DTrp) and first vertical derivative (DTrpVD). From these fields, the structural–geological cartography based on their regionalization is carried out, by distinguishing the areas of intense maximum values, maximums, intermediate values, minimums and very intense minimums. The intense maximum values, as a rule, respond to the presence of ophiolites and the very intense minimums and minimums, to the sediments of the basins. Tectonic alignments (faults) are determined from the total horizontal derivative of DTrp (DTrpTHD). Also, regional–residual separation is carried out from the AAC at 500 m, with the purpose of revealing weak magnetic targets (e.g., the rise of volcanic rocks over the Catalina oilfield). Finally, quantitative estimates of the depth to magnetic targets under the sediments are made from the magnetic field inclination derivative (TDR) (Pardo Echarte et al. 2019).

Non-conventional Geophysical–Geochemical Exploration Methods Used in Cuba.

1.2.2.4 Airborne Gamma Spectrometry

This method offers the potential to map and subdivide acid-medium and metamorphic igneous rocks and highlights rock types that are characterized by unusual amounts or very low proportions of radioelements such as basic–ultrabasic complexes. In less radioactive environments, such as volcano-sedimentary terrains and sedimentary basins, the most subtle contrasts also provide reliable mapping guides. The advantages of AGS are in comparison with other remote sensing techniques in the mapping of soil variations in areas of dense vegetation and areas of flat terrain. Regarding accumulations of oil and gas, the decomposition of clays in the soils as a result of the microseepage of light hydrocarbons is responsible for the minimum radiation observed. Finally, AGS data can complement the structural interpretation of other geophysical data, since they play an important role in the control of surface geology, where some structures that do not produce anomalous magnetic and gravity responses are deduced from these data (Pardo Echarte and Cobiella Reguera 2017).

In the AGS, the minimums of the K/Th relationship are mapped (fundamentally from the minimums of K) and the local maximums of U(Ra), in a majority way in its periphery, in order to indicate the localities linked with active zones of vertical microseepage of light hydrocarbons (Pardo Echarte et al. 2019). In addition, the known saline mapped structures are reflected by areas of increased K values. As a rule, the majority of the anomalous mapped localities, linked to presumed hydrocarbon microseepage, is located south of the Remedios TSU, which reinforces the criterion of interest for conventional oil and gas from the Camajuaní and Placetas TSUs.

1.2.2.5 Redox Complex

It is made up of a set of methods (used independently in the world with the same purpose): Redox Potential, Magnetic Susceptibility, Spectral Reflectance and Soil-geochemistry (variant of the metal mobile ion (MMI) method). The fundamentals of this complex of methods are included in a Patent (Pardo Echarte 2000) and in the Protected Work, Redox System (Rodríguez Morán 2005).

The Redox Complex is a complex of non-conventional geophysical–geochemical exploration techniques, used for the indirect detection and evaluation of various objects of a metallic nature (such as hydrocarbon accumulations: presence of V, Ni, Fe, Pb, Zn), which is based on the geochemical principle of vertical migration of metallic ions.

It is indicative of the physical–chemical processes and/or the modifications of the medium that take place in the upper part of the cut on the accumulations, conditioned by the diffusion haloes of light hydrocarbons and other satellite elements (metallic ions) that reach the surface.

Its objective is to establish the possible presence of hydrocarbons in the depth and its main characteristics (degree of preservation of the possible accumulation and estimates of the depth and quality of the hydrocarbon).

In Cuba, the application of these techniques has its antecedent in the works of Alfonso Roche and Pardo Echarte (1993) in which the possibility of remote mapping of potentially producing areas was established, revealed by minima of the K/Th relationship, and to verify the gas–oil nature of these anomalies by specifying their limits on land through the conjugate expression of minimums of the Redox Potential and maximums of the Magnetic Susceptibility of soils (Pardo Echarte and Rodríguez Morán 2016).

The exploration strategy of this complex of methods is based on considering a set of scenarios (structural maps by seismic, satellite Images, maps of potential fields, AGS and DEM) from which anomalous indications complexes are established (favorable areas resulting from the integrated prospective mapping) to which a cross-sectional reconnaissance profile is drawn for ground verification by the Redox Complex.

1.2.2.6 Integrated Prospective Mapping. Response Generalized Models (RGM) and Perspective Sectors

Integrated prospective mapping is carried out with the final purpose of reducing areas, based on considering a series of evaluation criteria (Pardo Echarte et al. 2019):

- Weak local gravimetric maxima, which reflect positive structures (due to the lifting of the denser carbonates and volcanics), within the limits of certain values of the reduced to the pole magnetic field;
- Weak residual magnetic maxima, related to the presence of diagenetic iron oxides and sulfides;
- Minimum values of the K/Th relationship, with local maxima of U(Ra), mostly in its periphery;
- Remote sensing anomalies;
- Positive residual geomorphic anomalies.

The interpretation focuses on the cartography of favorable areas made up of groups of possible hydrocarbon microseepage, within a specific geological–structural framework (mapped by gravimagnetometry), which have to be recognized and evaluated on land, later, by the Redox Complex.

The resulting total anomalous picture is verified with the Response Generalized Models (RGM) on the known accumulations (oilfields) previously determined, in order to establish the perspective sectors, on which the 2D defining seismic will be carried out (if there is none), prior to making decisions for exploratory drilling.

The most comprehensive RGM available to date are presented in Figs. 1.1, 1.2, 1.3 and 1.4.

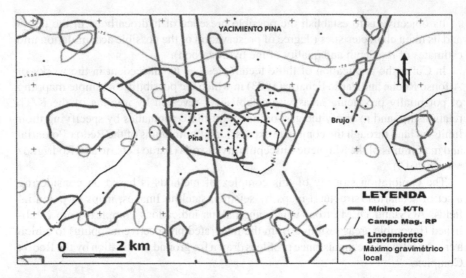

Fig. 1.1 RGM pattern of the Pina oilfield, Central Basin. In red, minimum of the K/Th ratio; in pink, local maxima of U(Ra); in black and green, contours of the RP magnetic field; in gray-green, tectonic alignments (faults) and local maximums by gravimetry; the black points correspond to oil wells (Pardo Echarte et al. 2019) (Color figure online)

1.3 Results and Discussion

As a result of a rigorous and exhaustive work of systematization–generalization of the investigations with NSNCEM in Cuba, a very small number (four) of perspective sectors have been established where, based on the information presented, the possible existence of hydrocarbon accumulations at depth is predicted. With the exception of a single sector (Maniabón), the results of a recognition profile by the Redox Complex are available, with relevant contribution from soil geochemistry (anomalous presence of elements V, Ni, Fe, Pb and Zn on the ground).

In general, the proportions of the areas of interest (Fig. 1.5) vary from very small (2–3 km^2, Motembo SE), small (3–5 or more square kilometers, Oeste de Ceballos and La Vigía, in Central Cuenca), to medium (50–70 km^2, Maniabón (Valdivia Tabares et al. 2015)). The quality of the hydrocarbons expected ranges from naphtha and light oil (Motembo SE), light oil (Oeste de Ceballos and La Vigía), to light–heavy oil (Maniabón), all consequences of dismigration and/or secondary migration processes. The depth of the predicted accumulations varies between 300 and 1200 m, which makes them very attractive from an operational point of view.

The validation of the perspective sectors by the 2D reflection seismic (mapped structures), it have for the sectors of Oeste de Ceballos (Martínez Rojas et al. 2006) and Maniabón (Valdivia Tabares et al. 2015) and/or by drilling wells, for Motembo SE and Maniabón. In the case of La Vigía, the realization of a 2D seismic profile (already planned) that will cross the revealed anomalous complex is pending, in order to validate it and propose, if positive, the location of an exploratory well.

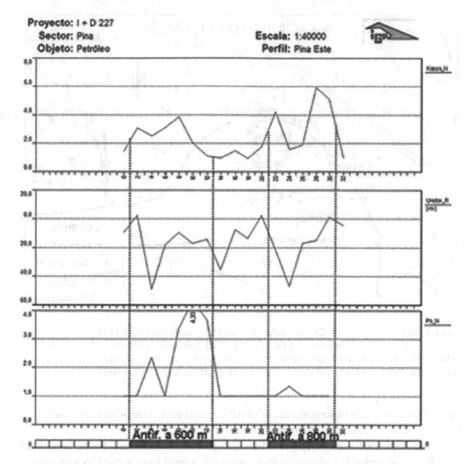

Fig. 1.2 Results of the recognition profile with the Redox Complex in the Pina oilfield. Positive anomalous results of the soil geochemistry (bar in red) on the AC showing the two anti-formic structures present; the distance between points is indicative (Pardo Echarte et al. 2019) (Color figure online)

A succinct argumentation of the four perspective sectors is presented, based on the anomalous NSNCEM complexes revealed, separated into Western Cuba (Motembo SE), Central Cuba (Oeste de Ceballos and La Vigía, in Central Basin) and Central-Eastern Cuba (Maniabón).

1.3.1   Western Cuba, Motembo SE Sector

The Motembo naphtha oilfield (Rodríguez and Kolesnikov 1970; Echevarría et al. 1991) was discovered in the year 1880–81. The maximum monthly production was about 26,000 barrels in November 1941. The naphtha accumulation zones correspond

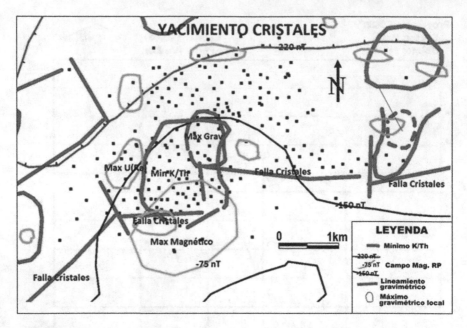

Fig. 1.3 RGM Pattern of the Cristales oilfield, Central Basin. In red (thick line), minimum of the K/Th ratio; in pink, local maxima of U(Ra); in red, black and green, contours of the RP magnetic field; in gray-green, tectonic alignments (faults) and local maximums by gravimetry; the black points correspond to oil wells (Morales González et al. 2020) (Color figure online)

to the fractured zones of serpentines, of small size and with little or no communication between them. Currently, the original reservoir is exhausted.

In the southeast sector (SE) of Motembo (limited to the east by an important N-NW course tectonic dislocation that apparently could have served as a migration route for hydrocarbons in the reservoir area), the following anomalous complex (AC) is observed, with an extension of 2–3 km^2 (Fig. 1.6):

- A minimum of the K/Th ratio (red line).
- A maximum of U(Ra) (pink line), coinciding with the minimum of the K/Th ratio.
- A remote sensing anomaly (RS), also coincidental (gray-green line).

It is worth noting, as an interest, the spatial position of this AC, close to the edge of the ophiolitic massif, as well as its uranium character, where a very probable relationship with hydrocarbons is revealed. Near the Motembo town (to the NW), there are two oil wells (Vesubio 24 (S/Autor 1954) and Motembo 1X (Sherritt 1995)). The first, very close to the referred AC, with a depth of 381 m, had naphtha inflows at 327 m and light oil at 342 m. The second, more distant from the AC, with a depth of 1941 m, with slight manifestations of gas.

The results of the recognition work by the Redox Complex (Fig. 1.7), according to a profile (blue line in Fig. 1.6), were abnormally positive according to the soil geochemistry at its western end (within the area of the AC-K/Th-U(Ra)-RS-), where

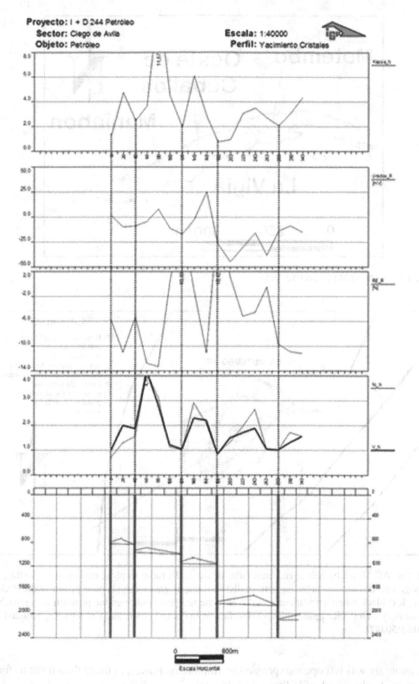

Fig. 1.4 Results of the recognition profile with the Redox Complex in the Cristales oilfield (quantitative interpretation). The distance between points is indicative (Pardo Echarte and Rodríguez Morán 2016)

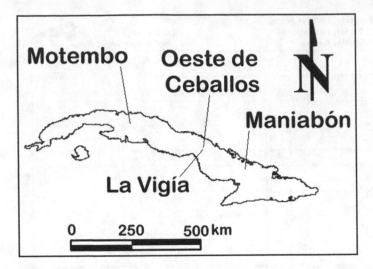

Fig. 1.5 Location of perspective sectors

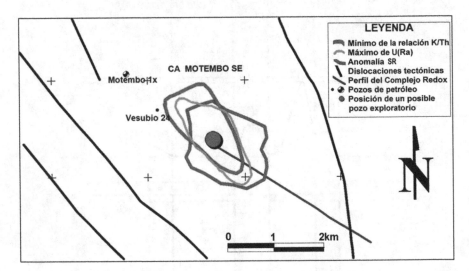

Fig. 1.6 AC Motembo SE: in red, minimum of the K/Th ratio; in pink, maximum of U(Ra); in gray-green, RS anomaly; in black, tectonic dislocations, by gravimetry; in blue, profile of the Redox Complex; black points correspond to oil wells; the red point indicates the position of a possible exploratory well (Color figure online) (Modified from Pardo Echarte and Cobiella Reguera 2017). Scale 1:50,000

the anomaly was left open; expressed by notable increases, in more than three to four times the background, of V, Pb and Zn. In addition, another narrow anomalous zone is

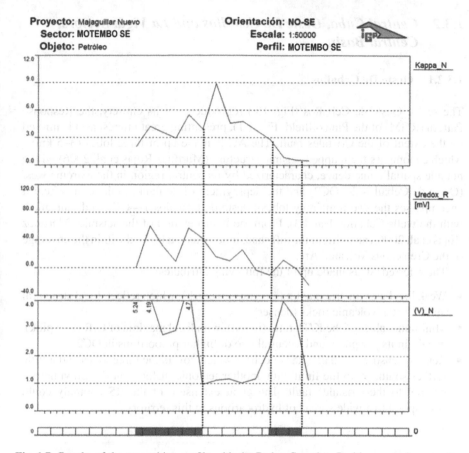

Fig. 1.7 Results of the recognition profile with the Redox Complex. Positive anomalous results of the Soil-geochemistry (bar in red) on the AC Motembo SE (western end of the profile, where the anomaly remains open) and the N-NW course tectonic dislocation (eastern end of the profile); the distance between points is indicative (Pardo Echarte and Cobiella Reguera 2017) (Color figure online)

observed at the eastern end of the profile, apparently linked to the aforementioned N-NW direction tectonic dislocation, which confirms its possible role as a hydrocarbon migration route in the area.

According to this information, the possible gas–oil target linked to the referred AC, with an area of 2–3 km^2, would be, shallow (300–400 m) in crushed ophiolites, covered by massive ophiolites. This represents an objective of little interest, due to its limited exploitable resources, but at the same time, of high interest, due to the operating facilities and the high quality of the expected hydrocarbons.

1.3.2 Central Cuba, Oeste De Ceballos and La Vigía Sectors, Central Basin

1.3.2.1 Oeste De Ceballos

The sector Oeste de Ceballos (Fig. 1.8) reproduces, in its entirety, the Response Pattern (RGM) of the Pina oilfield (Fig. 1.1), presenting great interest as it is limited by the extent of the Cristales fault. The AC is made up of three lobes (3–5 km^2), which encompass the mapped seismic structure (Martínez Rojas et al. 2006) with a notable spatial coincidence, characterized by its central region in the extreme west (Oeste de Ceballos 2—OC2) and the deployment of its flank, to the east, where it encompasses the two remaining lobes (Oeste de Ceballos 1—OC1 and, unnamed, with the wells Ceballos 1 and 3). From the interpretation of the seismic (Martínez Rojas et al. 2006), an uplift of the tuff sandstones was established within the coverage of the Cretaceous Volcanic Arc.

The referred AC is made up of the following attributes:

- Weak local gravimetric maxima (in gray-greenish), which reflect, local structural uplifts of the volcanic rocks (denser);
- Minimum values of the K/Th ratio (in red), with local maxima of U(Ra) (in pink), mostly in its periphery and a central one of higher proportions in OC2;
- Remote sensing anomaly (RS, in ocher) which covers the three lobes and a little further south, up to the limit with another tectonic dislocation (by gravimetry), parallel to the Cristales fault. The great extension of the RS anomaly could correspond to a wide spread of hydrocarbons in this territory.

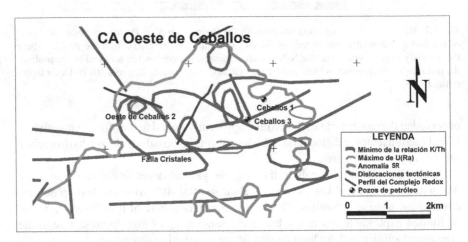

Fig. 1.8 AC Oeste de Ceballos: in red, minimum of the K/Th ratio; in pink, maximum of U(Ra); in gray-green, tectonic alignments and local gravimetric maxima; in ocher, RS anomaly; in blue, profile of the Redox Complex; the black point corresponds to the Ceballos 1 and 3 oil wells (color figure online) (Modified from Pardo Echarte et al. 2019)

The recognition profiles of the Redox Complex (blue line) on the OC2 and OC1 lobes are presented in Figs. 1.9 and 1.10, respectively (Pardo Echarte et al. 2019). According to their results:

- The possible presence of hydrocarbons in the depth of the OC2 lobe is confirmed, based on the existence of increases greater than or equal to three times the background of the correlated contents of V and Ni (hydrocarbon indicator elements), coinciding with increases of Fe, Pb and Zn. The results on the OC1 lobe are poorer, which means less interest.
- A correlation between the maximum of Magnetic Susceptibility with the increase of the chemical elements (V and Ni) is observed in OC2. The behavior of the Redox Potential is not diagnostic, as the maxima, associated with possible gaseous escapes, predominate.

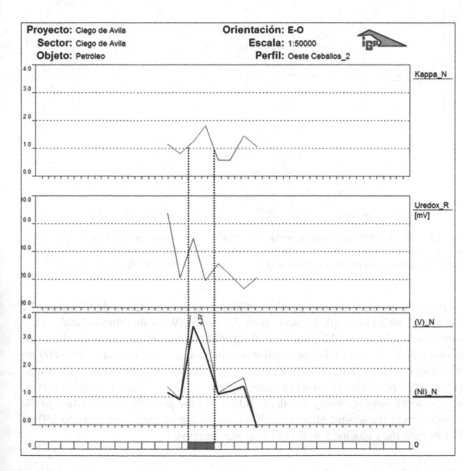

Fig. 1.9 Results of the recognition profile with the Redox Complex in the Oeste de Ceballos 2 sector. Positive anomalous results of the Soil-geochemistry (bar in red) on the AC. The distance between measurement points is indicative (Pardo Echarte et al. 2019) (Color figure online)

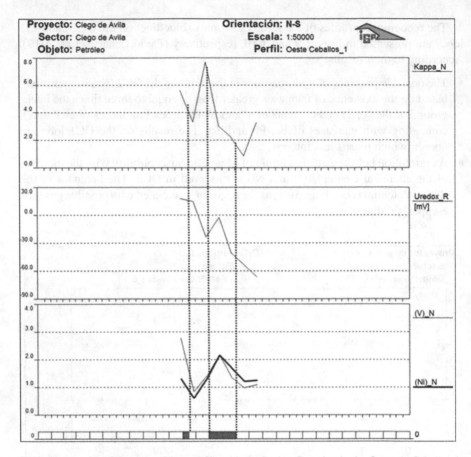

Fig. 1.10 Results of the recognition profile with the Redox Complex in the Oeste de Ceballos 1 sector. Anomalous poorer results of the soil geochemistry (bar in red) on the AC. The distance between measurement points is indicative (Pardo Echarte et al. 2019) (Color figure online)

From the point of view of another validation of the AC Oeste de Ceballos, the Ceballos 1 well (1450 m), located about 3.5 km S-SW of the Pina oilfield, in the innominate lobe of the AC, has several intervals of interest given by its shows of hydrocarbons. Presents oil impregnation in sandstones in the intervals (863–869 m; 877–886 m; 910–928 m; 940–949 m) and limestone oil impregnation in the range (1013–1015 m), although no oil or water input in all cases. This information serves, at least to fix, approximately, the depth interval of the possible accumulation in the central part of the seismic structure (OC2), which is predicted to be close to 850 m. For its part, the Ceballos 3 well (1512 m) was negative.

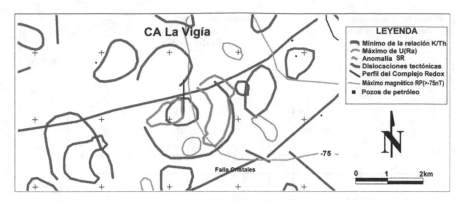

Fig. 1.11 AC La Vigía: in red, minimum of the K/Th ratio; in pink, maximum of U(Ra); in gray-green, tectonic alignments and local gravimetric maxima; in ocher, RS anomaly; in green, RP magnetic maximum (>−75 nT); in blue, profile of the Redox Complex; the black points correspond to oil wells (Color figure online) (Modified from Morales González et al. 2020)

1.3.2.2 La Vigía

La Vigía sector reproduces the RGM Pattern of the Cristales oilfield (Fig. 1.3) in its entirety, in addition to having a very interesting expression in the Redox Scenario (RS anomaly), which reveals the possible preservation of an accumulation (Fig. 1.11). According to the interpretation, the sector would have an extension somewhat greater than 5 km^2, being associated with a fault, presumably a feeder, to the north and parallel to the Cristales fault. The AC presents, in its southern portion, a local gravimetric maximum with an amplitude less than 0.5 m Gal and a slight increase to the north of it, coinciding with a local minimum of the K/Th relationship. This reveals the existence of a more sunken block to the north of the mentioned fault.

The RP magnetic field reveals a maximum of −13 nT of longitudinal heading, which covers the southern and northern part of the feeder fault, caused by the possible lifting of the top of the volcanics at a depth of approximately 600–800 m, according to the interpretation from the magnetic field inclination derivative (TDR), comparable with the 600–800 m found in Cristales for this same information and well data. The airborne gamma spectrometry from the K/Th relationship reveals two minima, with maxima of U(Ra) mainly in its periphery, one of greater proportions in the southern block and a smaller one in the northern block, both related to the same possible accumulation.

The results for the evaluation of the gas–oil nature through the Redox Complex in the AC La Vigía (Morales González et al. 2020) are presented in Fig. 1.12. These show the spatial correspondence of maxima (greater than 1.5 times the background value) of the Magnetic Susceptibility with minima (between −20 and −5 mV) of the Redox Potential and correlated increases of V and Ni (between 2 and greater than 3 times the background value), both for the south block as for the north (only one station). Such spatial correspondence makes it possible to establish the possible

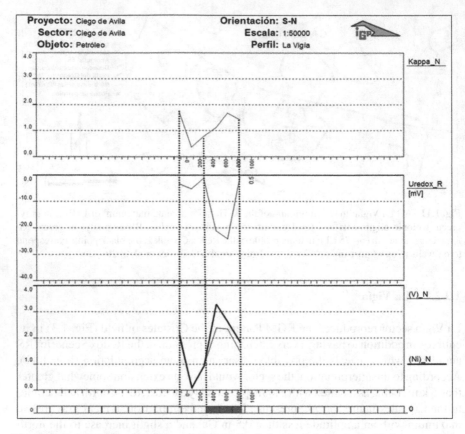

Fig. 1.12 Results of the recognition profile with the Redox Complex in the La Vigía sector. Positive anomalous results of the soil geochemistry (bar in red) on the AC. The distance between measurement points is indicative (Morales González et al. 2020) (Color figure online)

presence of a preserved accumulation of light hydrocarbons (given the minimum character of the Redox Potential and its value), in the order of depth previously proposed by the TDR (around 800 m).

Unfortunately, there is no seismic information and/or wells that support the interest of this sector, although a 2D seismic profile has already been planned that crosses, heading NE-SW, the referred AC on its southern flank.

1.3.3 Central-Eastern Cuba, Maniabón Sector

In the Maniabón sector, numerous reports on the existence of large superficial shows of hydrocarbons, dating from 1919 and where heavy oil was extracted from wells

5 m deep since colonial times (Linares et al. 2011), have encouraged the geologists for the exploration of oil and gas fields in the area.

There have been several exploratory attempts that have coincided in the area of Block 17: started in 1946 and 1947, the drilling of shallow wells Templanza 1 (495.3 m) and Fortaleza 1 (304.2 m), located in the vicinity of the Bernabé and La Anguila, respectively (areas of high impregnation on the Maniabón structure), where ophiolite productions were reported with an approximation of one barrel per day. Subsequently, to the north of the area of the shows, the Puerto Padre 1 well was drilled in 1958 (1099.2 m) and Farola Norte 1, in 1998 (2314 m), without productive results; until the recent drilling in 2011 of the Picanes 1X well (3568 m), toward the center of the block. The latter, in the final stages of drilling, showed considerable gas inputs to the mud, measuring in the order of 60% of the total gas (Pérez Martínez et al. 2013). The aforementioned aspects are irrefutable evidence of the existence of elements of an active oil system throughout the area.

It is distinguished in the town of Maniabón, with a notable spatial coincidence on the apical part of a large seismic structure ($50\text{-}70\text{km}^2$—Maniabón structure; Valdivia Tabares et al. 2015), the following AC (Fig. 1.13):

- Two minima of the K/Th ratio (in red).
- Maxima of U(Ra) (in pink) in the periphery of the previous minimums.
- A positive residual morphometric anomaly (0.4 m) (in black), coinciding with the minimum of the Fortaleza 1 well.

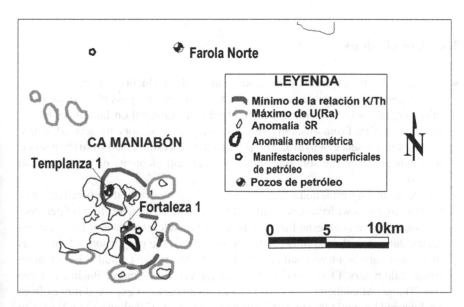

Fig. 1.13 AC Maniabón, northeastern region of Block 17. In red, minima of the K/Th ratio; in pink, peripheral maxima of U(Ra); in black, positive residual morphometric anomaly; in blue-thick line, RS anomalies and; in blue-thin line, north coast (NE); the black dots represent oil wells; small pentagons correspond to surface shows of hydrocarbons (Color figure online)

- Remote sensing anomalies (RS) (in blue).

According to Valdivia Tabares et al. (2015), the presence of abundant hydrocarbon emanations and areas with high impregnations concentrated in a large area indicate, by analogy with the main oilfields, the existence of an appreciable deposit in depth. Therefore, when taking into account the reservoir depth of the Maniabón target (slightly greater than 1000 m) and the secondary migration through subvertical fault systems, the total area of the possible deposit will include at least the area occupied by the superficial shows of hydrocarbons. The quality of the expected crude oil is always higher than that of the North Belt of Heavy Crudes, as there are Family III oils associated with the possible upper deposit (Maniabón).

According to the same authors, in the Maniabón area, as a result of the integration of conventional and non-conventional methods, the spatial correlation between the area occupied by surface oil emanations and the dimensions of the deposit was obtained (according to the charge model), as the radiometric anomalies (maxima of U(Ra)) are indicative of the southern edge of the possible accumulation. The vertical association of the seal break points with the two areas of abundant oil impregnation is assumed. These and other criteria constitute elements to propose the presence of an accumulation in depth, and therefore, justify the drilling of an exploratory well in the area, whose best location would be, in the opinion of the authors, in the north-central part of the Maniabón structure, right between the two areas of high oil impregnation (between the Templanza 1 and Fortaleza 1 wells).

1.4   Conclusions

- The non-seismic exploration methods, during the reduction of areas, lead to the mapping of favorable sectors, given by groups of possible hydrocarbon microseepage, which have to be recognized and evaluated on land, later by the Redox Complex. From this, only very few perspective sectors are derived, where the possible presence of hydrocarbons in the depth is established from the anomalous indications of the soil geochemistry (chemical elements of vanadium and nickel).
- As a result of a rigorous and exhaustive work of systematization–generalization of the investigations with these methods in Cuba, a reduced number (four) of perspective sectors have been established, where the possible existence of hydrocarbon accumulations in the depth is predicted. In general, the proportions of the areas of interest vary from very small (2–3 km^2, Motembo SE), small (3–5 or more square kilometers, Oeste de Ceballos and La Vigía, in Central Basin), to large (50–70 km^2, Maniabón). The quality of the hydrocarbons expected ranges from naphtha and light oil (Motembo SE), light oil (Oeste de Ceballos and La Vigía), to

light–heavy oil (Maniabón), results of dismigration processes and/or secondary migration. The depth of the predicted accumulations varies between 300 and 1200 m, for all cases, which makes them very attractive from an operational point of view.

- The validation of the perspective sectors by 2D reflection seismic (mapped structures) is available for the sectors of Oeste de Ceballos and Maniabón and, by drilling wells (presence of oil in the subsoil), for Motembo SE and Maniabón. In the case of La Vigía, the realization of a 2D seismic profile that will cross the revealed anomalous complex is pending in order to validate it and propose, if positive, the location of an exploratory well.

Acknowledgements The authors thank the Centro de Investigación del Petróleo and its Technical Archive for allowing the publication of non-confidential information on their research, as well as to Dr. Reinaldo Rojas Consuegra, to Dr. Juan Guillermo López Rivera, to Dr. Evelio Linares Cala and M.Sc. Orelvis Delgado López, researchers at this institution, for the exhaustive and rigorous review of the manuscript.

References

Alfonso Roche JR, Pardo Echarte ME (1993) Informe sobre los trabajos metodológicos–experimentales de métodos geofísicos y geoquímicos no convencionales para la prospección de hidrocarburos someros en Cuba Septentrional (Unpublished report). ENG, La Habana.

Colectivo de Autores (2008) Mapa Digital de las Manifestaciones de Hidrocarburos de la República de Cuba a escala 1:250,000. Unpublished report. Centro de Investigaciones del Petróleo, La Habana

Colectivo de Autores (2009) Mapa Digital de los Pozos Petroleros de la República de Cuba a escala 1:250,000. Unpublished report. Centro de Investigaciones del Petróleo, La Habana

Colectivo de Autores (2010) Mapa Geológico Digital de la República de Cuba a escala 1:100,000. Unpublished report. Instituto de Geología y Paleontología, Servicio Geológico de Cuba, La Habana, Cuba

Echevarría G, Hernández Pérez G, López Quintero JO, López Rivera JG, Rodríguez Hernández R, Sánchez Arango J, Socorro Trujillo R, Tenreyro Pérez R, Iparraguirre Peña JL (1991) Oil and gas exploration in Cuba. J Petrol Geol 14(3):259–274

Jiménez de la Fuente L, Pardo Echarte ME (2020) Escenarios REDOX de sectores perspectivos en Cuba. Unpublished report. Centro de Investigación del Petróleo, La Habana, p 7

Linares E, García Delgado D, Delgado López O, López Rivera JG, Strazhevich V (2011) Yacimientos y manifestaciones de hidrocarburos de la República de Cuba. Centro de Investigaciones del Petróleo, La Habana, 480 p

Martínez Rojas E, Toucet Téllez S, Sterling Baños N, Iparraguirre Peña JL (2006) Informe sobre la reinterpretación geólogo-geofísica y evaluación estructural del Bloque 21. (Reinterpretación sísmica terrestre 2D). Unpublished report. Centro de Investigaciones del Petróleo (CEINPET), La Habana, Cuba, 38 p

Mondelo Diez F, Sánchez Cruz R et al (2011) Mapas geofísicos regionales de gravimetría, magnetometría, intensidad y espectrometría gamma de la República de Cuba, escalas 1:2,000,000 hasta 1:50,000. Unpublished report. IGP, La Habana, 278 p

Morales González J, Rodríguez Morán O, Pardo Echarte ME (2020) Possible gaso-petroleum occurrence from non-seismic and non-conventional exploration methods in the Central Basin, Cuba. Boletín Ciencias de la Tierra, No 47, pp 15–20

Pardo Echarte ME (2000) Certificado de Autor de Invención Nro. 22 635, concedido por la Resolución Nro. 475/2000, sobre: "Método de medición del Potencial Redox en suelos y su aplicación combinada con la Kappametría a los fines de la prospección geológica". Clasificación Internacional de Patentes: G01V 3/00. Oficina Cubana de la Propiedad Industrial, Ciudad de la Habana, 2 p

Pardo Echarte ME, Rodríguez Morán O (2016) Unconventional methods for oil & gas exploration in Cuba. Springer Briefs Earth Syst Sci. https://doi.org/10.1007/978-3-319-28017-2

Pardo Echarte ME, Cobiella Reguera JL (2017) Oil and gas exploration in Cuba: geological-structural cartography using potential fields and airborne gamma spectrometry. Springer Briefs in Earth Syst Sci. https://doi.org/10.1007/978-3-319-56744-0

Pardo Echarte ME, Rodríguez Morán O, Delgado López O (2019) Non-seismic and non-conventional exploration methods for oil and gas in Cuba. Springer Briefs Earth Syst Sci. https://doi.org/10.1007/978-3-030-15824-8

Pérez Martínez Y et al. (2013) Proyecto 7054. Etapa 1.4. Informe final sobre fundamentación de pozo en el Bloque 13. Archivo Técnico CEINPET, La Habana, 21 p

Price LC (1985) A critical overview of and proposed working model for hydrocarbon microseepage. US Department of the Interior Geological Survey. Open-File Report 85-271

Rodríguez R, Kolesnikov L (1970) Informe sobre el área de Motembo y Corralillo. Unpublished report. Archivo CEINPET O-45, La Habana, 53 p

Rodríguez Morán O (2005) Certificación de Depósito Legal Facultativo de Obras Protegidas, Registro Nro. 1589-2005, sobre Sistema Redox (*Software*), in CENDA. Ciudad de la Habana, 1 p

Sánchez Cruz R, Mondelo Diez F et al. (2015) Mapas Morfométricos de la República de Cuba para las Escalas 1:1,000,000–1:50,000 como apoyo a la Interpretación Geofísica. Memorias VI Convención Cubana de Ciencias de la Tierra, VIII Congreso Cubano de Geofísica. http://www.cgiar-csi.org/data/srtm-90m-digital-elevation

S/Autor (1954) Columnas lito paleontológicas de los pozos Vesubio 24, 25 y 26. Unpublished report. Archivo CEINPET 91950, La Habana, 3 p

Saunders DF, Burson KR, Thompson CK (1999) Model for hydrocarbon microseepage and related near-surface alterations. AAPG Bulletin 83(1):170–185

Sherritt (1995) Informe sobre el pozo Motembo 1X. Unpublished report. Archivo CEINPET E-260, La Habana, 7 p

Schumacher D (1996) Hydrocarbon-induced alteration of soils and sediments. In: Schumacher D, Abrams MA (eds) Hydrocarbon migration and its near-surface expression: AAPG Memoir 66, pp 71–89

Schumacher D (2014) Minimizing exploration risk: the impact of hydrocarbon detection surveys for distinguishing traps with hydrocarbons from uncharged traps. GeoConvention 2014: FOCUS

Valdivia Tabares CM, Veiga Bravo C, Martínez Rojas E, Delgado López O, Domínguez Sardiñas Z, Pardo Echarte ME et al (2015) Informe de resultados de la evaluación del potencial de hidrocarburos del Bloque 17. Unpublished report. Archivo CEINPET, La Habana, p 116

Chapter 2
Geological–Structural Mapping and Favorable Sectors for Oil and Gas in Western Cuba Through Non-seismic Exploration Methods

Manuel Enrique Pardo Echarte, Lourdes Jiménez de la Fuente, and Yeniley Fajardo Fernández

Abstract It is known that non-seismic exploration methods offer necessary and important information on the geological–structural mapping of the territories and on the presence in them of vertical areas of active microseepage of light hydrocarbons, witnesses to possible accumulations at depth. That is why the benefits of using these methods, prior to their integration with geological and seismic data, translate into a first approximation, valid for an initial understanding of geology and mapping of favorable areas of possible gas–oil interest. Such are the objectives of the investigation at the western Cuba region (land exploration blocks 6, 7, 8A and 9A). To meet these objectives, gravimetry and aeromagnetometry at a scale of 1:50,000 and 1:250,000, AGS at a scale of 1:100,000, (ASTER satellite images) and the Digital Elevation Model 90 × 90 m of the territory were processed. For the geological interpretation, the Digital Geological Map of the Republic of Cuba at a scale of 1:100,000 was used. The geological–structural cartography of the study region, based on the gravimagnetic and morphometric data, establishes possible structural highs which alternate with depressed areas. A version of favorable sectors mapping of gas–oil interest (linked to conventional oil and gas from the Camajuaní and Placetas Tectonostratigraphic Units) is based on the presence of a complex of geophysical–morphometric indicator anomalies. It considers: minimums of the K/Th ratio and the local maxima of U(Ra) in its periphery; remote sensing anomalies as well as; local gravimetric, magnetic and morphometric maxima. The work presents an account of the processing and interpretation of non-seismic exploration methods and, as practical results, the foundation of the main favorable sectors of possible interest for exploration. The

M. E. Pardo Echarte (✉) · L. Jiménez de la Fuente · Y. Fajardo Fernández
Centro de Investigación del Petróleo, Churruca, No. 481, Vía Blanca y Washington, 12000 El Cerro, La Habana, CP, Cuba
e-mail: pardo@ceinpet.cupet.cu

L. Jiménez de la Fuente
e-mail: lourdes@ceinpet.cupet.cu

Y. Fajardo Fernández
e-mail: yeniley@ceinpet.cupet.cu

© The Author(s), under exclusive license to Springer Nature Switzerland AG 2022
M. E. Pardo Echarte et al., *Geological-Structural Mapping and Favorable Sectors for Oil and Gas in Cuba*, SpringerBriefs in Earth System Sciences,
https://doi.org/10.1007/978-3-030-92975-6_2

geological–structural cartography of the study region, based on gravimagnetic and morphometric data, allowed to clarify the structural picture of the territory, where geological structures of Cuban direction (SE-NW course) and others of SW-NE and latitudinal (EW) course predominate.

Keywords Non-seismic exploration methods · Remote sensing · Gravimetry · Aeromagnetometry · Airborne gamma spectrometry · Morphometry · Favorable sectors for hydrocarbons · Cuba

Abbreviations

AGS	Airborne gamma spectrometry
TSU	Tectonostratigraphic unit
Fm	Geological formation
NACM	North American continental margin
OA	Ophiolite association
CVA	Cretaceous volcanic arcs
CNTFB	Cuban North thrusted and folded belt
DEM	Digital elevation model
Gb	Gravimetry
DT	Aeromagnetometry
RS	Remote sensing
It	Total gamma intensity
AAC	Ascending analytical continuation
GbVD	First vertical derivative of gravity field
Gbres500	Gravity field, residual component
GbTHD	Total horizontal derivative of gravity field
DTrp	Reduced to the pole magnetic field, 1:50,000 scale
DTrpTHD	Total horizontal derivative of reduced to the pole magnetic field, 1:50,000 scale
DTrpreg500	Reduced to the Pole Magnetic Field, regional component, 1:50,000 scale
DT250rp	Reduced to the Pole Magnetic Field, 1:250,000 scale
DT250rpVD	First vertical derivative of reduced to the pole magnetic field, 1:250,000 scale
DT250rpTHD	Total horizontal derivative of reduced to the pole magnetic field, 1:250,000 scale
TDR	Magnetic field inclination derivative
DEMTHD	Total horizontal derivative of DEM
DEMres500	DEM, residual component
DEMreg500	DEM, regional component
CNOB	Cuban North oil belt

2.1 Introduction

It is known (Pardo Echarte et al. 2019) that non-seismic exploration methods offer necessary and important information on the geological–structural mapping of the territories and on the presence in them of vertical zones of active microseepage of light hydrocarbons, witnesses of possible accumulations in depth. That is why the benefits in the use of these methods, prior to their integration with geological and seismic data, are translated into a first approximation, valid for an initial understanding of geology and the mapping of areas of possible oil and gas interest; such are the objectives of this research.

The non-seismic exploration methods considered in this work are: remote sensing, gravimetry, aeromagnetometry, airborne gamma spectrometry (AGS) and morphometry. The penultimate also classifies as a non-conventional geophysical–geochemical method. The referred complex, with the exception of gravimetry, is based on the anomalous physical–chemical response of the medium in front of the hydrocarbon microseepage on the accumulations in depth and has for purposes the reduction of areas and the increase of the effectiveness of the exploration, with a more sensible reduction of its risks.

As a background to this work, there are the Research Report (unpublished) corresponding to the results of the application of non-seismic exploration methods in Block 6 (Pardo Echarte 2019) and the study carried out in Habana–Matanzas region (Pardo Echarte et al. 2019).

In the work, the geographical location, the general aspects of the regional geology of the territory, the geological task, the physical–chemical–geological premises that support the application of non-seismic exploration methods, and the materials and methods used, are exposed. After to a brief account of the processing and interpretation of non-seismic exploration methods, the research addresses, in the first instance, the results of the geological–structural mapping of the territory from gravimetric and magnetic data and then focuses on the results of the integrated prospective mapping, which leads to favorable areas for hydrocarbons.

2.1.1 Geographical Location

2.1.1.1 Land Exploratory Block 6

Block 6, totally terrestrial, is located in the northwestern region of Cuba (Fig. 2.1), encompassing the provinces of La Habana and Pinar del Río with an approximate area of 1400 km^2.

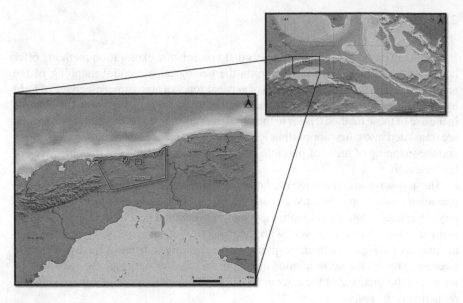

Fig. 2.1 Geographical location of Block 6

2.1.2 Land Exploratory Blocks 7, 8A and 9A

Block 7, entirely terrestrial, is located in the western region of Cuba, partially comprised of the provinces of La Habana, Mayabeque and Matanzas, with an approximate area of 1850 km². It is bounded by blocks 9A (to the east) and 8A (to the south), to the west, it borders a non-bid area (Fig. 2.2).

2.1.3 Regional Geological Model

2.1.3.1 Land Exploratory Block 6

According to Domínguez Gómez et al. (2005), the geological model is characterized by the presence of several mantles of the Sierra del Rosario Tectonostratigraphic Unit (TSU), which have been cut by several wells drilled in the region. It is assumed that underneath the rock stacks of the Rosario TSU are tectonic mantles of the Placetas TSU. In the composition of the section of this region, it is necessary to consider the important role played by the synorogenic deposits of the Campanian–Maastrichtian Upper Cretaceous of the Vía Blanca Formation (Fm.) and those of the Cretaceous–Paleogene limit of the Cacarajícara Fm. (according to Tada et al. 2003; Goto et al. 2008), as well as from the Paleocene–Lower and Middle Eocene, whose fundamental unit is the Manacas Fm., on top of a melánge that contains large and small blocks

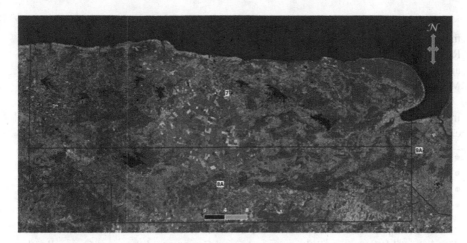

Fig. 2.2 Geographical location of the study area in a Google image. The limits of the land exploration blocks are outlined in red (Color figure online)

of ophiolites. These lithological complexes are included in the tectonic scales, as demonstrated by the Mariel 1-X well, and generate reflections through which it is possible to draw the structure of the folds. All sequences are deformed according to thrust tectonics; however, it must consider the role of neo-tectonic forces in the redesign of anticline zones such as Martín Mesa.

These deformations occurred, perhaps, as a consequence of the dislocations that occurred after the thrust phase. What stands out the most from the regional geological model of Block 6 is the presence of structural elevations, which are composed of piles of synorogenic rock scales and carbonates from the Sierra del Rosario TSU. These highs form chains or fringes in the WSW–ENE direction and alternate with structurally depressed zones, where the wells report large thicknesses of blocks of serpentines and volcanites included in a melánge sequence.

2.1.3.2 Land Exploratory Blocks 7, 8A and 9A

According to Pardo Echarte and Cobiella Reguera (2017), in the folded basement of Cuba, the pre-Cenozoic floor is formed by three complexes of different nature:

- Mesozoic passive continental paleomargin
- Mesozoic ophiolitic association
- Cretaceous volcanic arches (including its metamorphic basement and the Campanian–Maastrichtian sedimentary cover).

The passive Mesozoic continental paleomargin considers: a northern distensive margin—North American Continental Margin (NACM), extended between Pinar del Río and NW of Holguín, with a small area in the extreme eastern part of Cuba (Maisí); and a southern distensive continental margin with two areas—Isla de la Juventud and Macizo Escambray.

In the NACM, in the cuts between La Habana and Camagüey, on the surface and in the subsoil, from north to south, the following tectonostratigraphic units (TSU) can be distinguished:

- Cayo Coco
- Remedios
- Camajuaní
- Placetas.

The Camajuaní and Placetas units are detached from their basement, while Remedios is possibly para-autochthonous and Cayo Coco, autochthonous. Generally, the rocks of the ophiolitic association are structurally arranged on top of the Placetas unit, which contains the layers originally deposited further south.

The Mesozoic ophiolitic association (OA), in the northern ophiolitic belt, is formed by rocks of the oceanic lithosphere tectonically located on the NACM. Its rocks are represented by serpentinized ultramaphytes, serpentinites, mafic–ultramafic cumulative complexes and mafic rocks (intrusive and volcanic).

Regarding the volcano-sedimentary sequences of the OA, it is sometimes difficult to separate them from those of the Cretaceous Volcanic Arc (lower part?), for which detailed petrochemical and petrographic studies are required.

Between Pinar del Río and Camagüey, ophiolitic rocks underlie the Cretaceous volcano-sedimentary successions. The contact between the two is always tectonic. The latter contain a chaotic mixture of serpentinites and gabbroids with rocks from the aforementioned successions. In fact, the deformations and tectonic mixing of lithologies are so remarkable that, in essence, the belt is a great melánge.

In much of Cuba, structurally located on ophiolitic rocks and occupying, in general, a more southern position, the Cretaceous volcanic (insular) arches (CVA) are arranged, formed by volcanic and volcanic-sedimentary Cretaceous cuts, as well as its metamorphic substrate and a sedimentary cover from the Late Upper Cretaceous. In western Cuba, the outcrop of the Cretaceous volcano-sedimentary cuts is much more limited than in Central Cuba. The Lower Cretaceous rocks are represented by the Chirino Formation (Ducloz 1960), which, as in central Cuba, contains little sedimentary material. The Upper Cretaceous cut is of limited thickness and its volcanites are calcoalkaline, and contains abundant sedimentary intercalations. The integration of the OA and the CVA was called Zaza Terrain (Hatten et al. 1988).

According to Pardo Echarte and Cobiella Reguera (2017), regarding the synorogenic basins, along the north of Cuba, from the NW of Pinar del Río to Gibara (Holguín), the rocks of the NACM are covered by the deposits of the foreland basin. These are accumulated successions in front of the thrust mantles generated during the Cuban orogenesis, as a consequence of the erosion of its frontal region and the rapid subsidence of the basin, due to the weight of the thrust mantles. The sedimentation in these depressions is contemporaneous with the orogenic deformations and the dating of their deposits marks the age of the event (late Campanian–Maastrichtian to Paleocene–Lower Eocene).

There is a close overlap between the tectonic scales of the southern portion of the foreland basin, formed mainly by olistostromes and the scales of ophiolitic rocks, the Cretaceous Volcanic Arc and the NACM. This scaled belt is a folded and faulted belt, with alpine tectonics of fine scales, originated by a combination of compressional and gravitational tectonics (Cuban North Thrusted and Folded Belt-CNTFB).

According to Colectivo de Autores (2009b), the CNTFB is characterized by several levels of ramp folds against reverse faults of the NACM and its coverage. These folds have probably been further complicated by shear accidents. The deformed rocks span an age range from the Jurassic to the Eocene. Stacking of various ramp anticline folds is one of the main exploratory targets in the study region. These make up antiforms that are mappable with great difficulty by seismic techniques. The poor image obtained is the main obstacle to the development of exploratory work; only the one directly related to the envelope of the scale folds is observed as a horizon with high dynamic definition.

On the other hand, in the territory, the reservoirs, as a rule, are represented by intensely fractured and leached limestones, covered by a seal of clays from the Paleocene to the Eocene and, on occasions, by fractured serpentinites (Colectivo de Autores 2009b).

2.1.4   Geological Task

The geological task posed to the processing and geophysical–geochemical–morpho-metric interpretation of the study region, and the general objective of the research, consists of establishing favorable sectors linked to conventional oil and gas of the Camajuaní and Placetas TSUs, based on the presence of a complex of remote sensing, airborne gamma spectrometry, gravimetric, magnetic and morphometric indicator anomalies.

As a specific objective that precedes the previous one, it is proposed to carry out the geological–structural mapping by gravimagnetic and morphometric data of the territory. To meet these objectives, the following are processed: ASTER satellite images (RS) (Jiménez de la Fuente 2017), gravimetry (Gb) and aeromagnetometry (DT) at scales 1:50,000 and 1:250,000, airborne gamma spectrometry (AGS) at 1:100,000 scale, the Digital Elevation Model (DEM) 90 × 90 m and the petroleum information of the territory. For the geological interpretation, the Digital Geological Map of the Republic of Cuba at 1:100,000 scale (Colectivo de Autores 2010) was used.

2.1.5 Physical–Chemical–Geological Premises

The high density of carbonate, volcanic and ophiolitic rocks makes it possible to distinguish by gravimetric maximums their structural elevations. Likewise, the high magnetic susceptibility of volcanics and ophiolites allows them to be mapped by aeromagnetometry.

From the point of view of Surface Geochemistry, according to Price (1985), Schumacher (1996), Saunders et al. (1999) and Pardo and Rodríguez (2016), the Physical–Chemical–Geological Premises that support the application of non-conventional geophysical–geochemical–morphometric exploration methods are the following:

As the light hydrocarbons rise from the accumulation, bacterial oxidation produces, as a by-product, carbonic and organic acid, as well as hydrogen sulfide. For its part, carbonic acid reacts with clay minerals, destroying them, while creating secondary carbonate mineralization and silicification. Close to the surface, both materials are denser and more resistant to erosion, with an effect in the increase of the seismic velocity on the accumulation, as well as in the formation of erosional topographic maxima and resistivity maxima.

Regarding oil and gas accumulations, the decomposition of clays in soils as a result of the microseepage of light hydrocarbons is responsible for the minimum radiation observed on oil and gas fields: Potassium is leached from the system toward the edges of the vertical projection of the accumulation, where it precipitates, resulting in a "halo" of high values. Thorium remains relatively fixed, in its original distribution within insoluble heavy minerals, hence minimums of the K/Th ratio surrounded by maximums are observed on these deposits. In a majority way, in the periphery of these anomalies, local increases of U(Ra) are also observed.

Additionally, the aforementioned K/Th ratio offers the opportunity to eliminate a series of undesirable effects on spectrometric measurements (influence of lithology, humidity, vegetation and measurement geometry).

Regarding the role of hydrogen sulfide, its own presence conditions the formation of a reducing environment column (minimum of the Redox Potential) on the accumulation. This reducing environment favors, in turn, the conversion of non-magnetic iron minerals into more stable (diagenetic) magnetic varieties such as magnetite, maghemite, pyrrhotin and greigite, all responsible for the increase in the Magnetic Susceptibility of rocks and soils, as well as the presence of subtle local magnetic maxima on the accumulation. This fact explains the observed inverse correlation between both attributes (minimums of Redox Potential and maximums of Magnetic Susceptibility) and justifies the integration of the methods.

The arrival to the surface of the metallic ions contained in the hydrocarbons (V, Ni, Fe, Pb and Zn among others) conditions the presence of a subtle anomaly of these elements in the soil and a slight change in its coloration (darkening), which is reflected by Reduced Spectral Reflectance anomalies (minima) in rock and soil samples, as well as in satellite images (Redox Scenarios); these evidences also justify the integration of these techniques (Pardo Echarte and Rodríguez Morán 2016).

2.2   Materials and Methods

2.2.1   Information and Its Sources

The materials used and their sources are the following:

- ASTER satellite images of the region and their interpretation (Jiménez de la Fuente 2017).
- Gravimetric and magnetic field grids at scales 1:50,000 and 1:250,000, and of airborne gamma spectrometry (channels: It, U, Th and K) at scale 1:100,000 of the Republic of Cuba (Mondelo Diez et al. 2011).
- The DEM (90 × 90 m) used in this work was taken from Sánchez Cruz et al. (2015), with source at: http://www.cgiar-csi.org/data/srtm-90m-digital-elevation.
- Digital Maps of the Hydrocarbon Manifestations and the Oil Wells of the Republic of Cuba at a scale of 1:250,000 (Colectivo de Autores 2008 and 2009a, respectively).
- Digital Geological Map of the Republic of Cuba at 1:100,000 scale (Colectivo de Autores 2010).
- Petroleum information of the territory.

2.2.2   Methods and Techniques

The non-seismic exploration methods used in the work are:

- Remote sensing (RS)
- Gravimetry (Gb)
- Aeromagnetometry (DT)
- Airborne gamma spectrometry (AGS)
- Non-conventional morphometry.

The geophysical information processing was carried out with the Oasis Montaj-GeoSoft version 7.01 software.

2.3   Results

2.3.1   Information Processing and Interpretation

2.3.1.1   Land Exploratory Block 6

Gravimetry
The gravimetric field (Bouguer Reduction, 2.3 t/m^3) was subjected to a regional–residual separation from the Ascending Analytical Continuation (AAC) for the

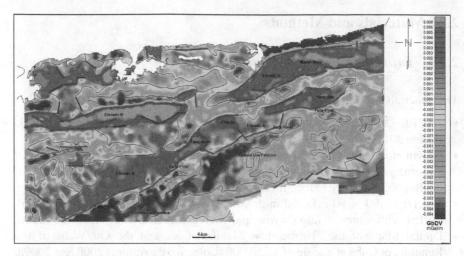

Fig. 2.3 Gravimetric mapping, from the GbVD field. In black thin line, limit of regionalization; in gray-greenish, tectonic alignment by gravimetry (Color figure online)

heights of 500, 2000 and 6000 m, given by the order of depth of the possible gas–oil objectives and the seismic study. For gravimetric mapping (Fig. 2.3), the first vertical derivative (GbVD), equivalent to a residual at 500 m, and the total horizontal derivative (GbTHD), was used to draw the tectonic alignments. In this figure, the maximums (possible structural elevations) are associated with the presence of stacks of non-orogenic carbonate rock scales from the Sierra del Rosario TSU and, the minimums and/or the non-anomalous field, with possible structural depressions.

Aeromagnetometry

The lithology of magnetic character (volcanic + ophiolites from Zaza Terrain) can be distinguished directly on the basis of observations of the reduced to the pole magnetic field (DTrp).

A regional–residual separation from the AAC at 500 m (Fig. 2.4) was carried out, and the tectonic alignments were drawn from the total horizontal derivative (DTrpTHD). In this figure, the maximums are associated with the presence of the Zaza Terrain (volcanic + ophiolites) and/or only by the CVA (S-SE) and; the minimums and/or the non-anomalous field, with possible structural depressions and/or the presence of stacks of non-orogenic carbonate rock scales from the Sierra del Rosario TSU, not magnetic.

An estimate of the depth of two magnetic targets was made to the SE of the Pinar Fault (within the CVA), from the of the magnetic field inclination derivative (TDR), yielding values between 1000 and 1300 m, respectively.

Airborne Gamma Spectrometry

For the AGS, the K/Th ratio was determined, in order to recognize the minima, presumably linked to active zones of vertical microseepage of light hydrocarbons.

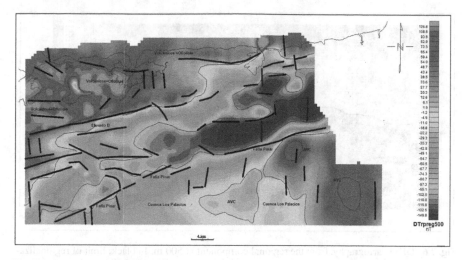

Fig. 2.4 Magnetic mapping, from the regional component DTrpreg500 field. In black thin line, limit of regionalization; in black thick line, tectonic alignment by aeromagnetometry (Color figure online)

The results of the AGS mapping (minimums of the K/Th ratio and local maximums of U(Ra) in its periphery), together with those of the geological–structural gravimetric regionalization and local gravimetric and morphometric maximums, are presented in Fig. 2.5 (Integrated prospective mapping—Favorable Areas).

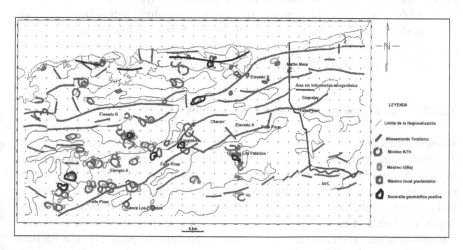

Fig. 2.5 Results of integrated prospective mapping (Favorable Areas). The blue line on the map is the boundary of the non-surveyed area. In black thin line, limit of regionalization; in red, minimum of the K/Th ratio; in pink, maximum of U(Ra); in gray-greenish, tectonic alignments and local gravimetric maxima; in black thick line, morphometric maxima (Color figure online)

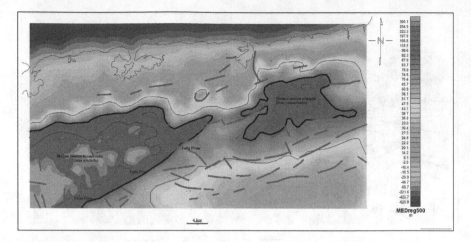

Fig. 2.6 DEM cartography, from the regional component at 500 m. In black, limit of regionalization; in red, tectonic alignment by morphometry. The morphometric data indicates the existence of a possible less conserved block (lower erosion cut) at the WSW of the Cayajabos structure (left block) (Color figure online)

Non-conventional Morphometry

The DEM (90 × 90 m) was subjected to regional–residual separation from the AAC at 500 m, according to the author's experience, with the purpose of regionalizing the relief field and tracing the tectonic alignments from the total horizontal derivative DEMTHD (Fig. 2.6). Local maxima are also determined (from the residual component DEMres500), possibly related to the processes of slight subsurface carbonatization and silicification that take place on the active vertical microseepage of light hydrocarbons. These local positive geomorphic anomalies are represented on the Favorable Areas map, together with the local gravimetric maxima (Fig. 2.5).

2.3.1.2   Land Exploratory Blocks 7, 8A and 9A

Remote Sensing

The studied area (Jiménez de la Fuente 2017) includes the territory between the towns of Guanabo and Seboruco, and between the coast and the Y-338000 coordinate. The processing consisted of the construction of band ratios that responded to lithological and mineralogical changes, produced by possible accumulations of hydrocarbons in depth. The ratios of bands 2/1 and 4/9 were constructed that allowed mapping areas with possible alterations by ferric oxides and carbonates, respectively. The RS limit of the outcropping carbonates is also indicated.

The interpretation criteria followed consider the division of the study region into two zones, North and South, where the areas corresponding to the following criteria were mapped:

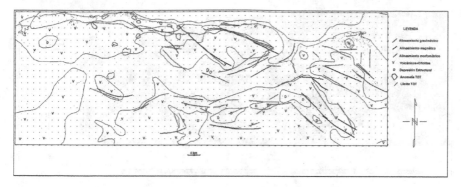

Fig. 2.7 RS anomalies, in the framework of gravimetric mapping and tectonic alignments. In brown, RS anomalies and RS limit; in black thin line, limit of regionalization; V, volcanic + ophiolites; D, structural depression; in gray-greenish, gravimetric alignment; in black, magnetic alignment; in blue, morphometric alignment

- North Zone (related to the main oilfields): Maximum values of carbonate ratio and high values of the iron ratio not related to anthropic elements.
- South Zone: Low-medium carbonate ratio values and high iron ratio values not related to anthropic elements.

The work does not perform the analysis of satellite images from the point of view of Reduced Spectral Reflectance (Redox Scenarios).

The results of the RS mapping are presented in Fig. 2.7, within the framework of gravimetric mapping and tectonic alignments.

Gravimetry

The gravimetric field (Bouguer Reduction, 2.3 t/m^3) was subjected to a regional–residual separation from the Ascending Analytical Continuation (AAC) for the heights of 500, 2000 and 6000 m, given by the order of depth of the possible gas–oil and seismic study objectives. For the gravimetric geological–structural mapping (Fig. 2.8), the first vertical derivative (GbVD) was used, equivalent to the residual at 500 m and the total horizontal derivative (GbTHD) for the tracing of the tectonic alignments. In this figure, the maximums are associated with the presence of volcanic rocks and ophiolites of the Zaza Terrain, and the minimums and/or the non-anomalous field, with structural depressions and carbonates of the North American Continental Margin (NACM). The residual field at 500 m (Gbres500) and the GbDV allow the mapping of very subtle local gravimetric maxima, associated with possible structural uplifts of carbonates and/or volcanics, with gas–oil interest.

Aeromagnetometry

Magnetic lithology (volcanic + ophiolites) can be distinguished directly on the basis of observations of the reduced to the pole magnetic field (DTrp and DT250rp). The magnetic field was also subjected to the first vertical derivative and the tectonic alignments are drawn from the total horizontal derivative DT250rpTHD field. The

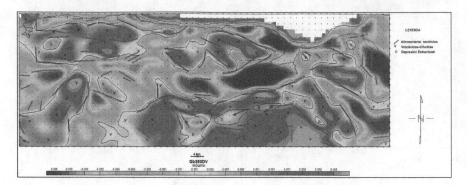

Fig. 2.8 Geological–structural mapping from gravimetric data. In black thin line, limit of regionalization; in gray-greenish line, tectonic alignments by gravimetry. The maximums are associated with the presence of volcanic rocks and ophiolites of the Zaza Terrain (V) and, the minimums and/or the non-anomalous field, with structural depressions and carbonates of the NACM (D) (Color figure online)

geological–structural cartography based on magnetic data is presented in Fig. 2.9. In this figure, the maximums are associated with the presence of volcanics and ophiolites from the Zaza Terrain, and the minimums and/or the non-anomalous field, to structural depressions and carbonates from NACM. The quantitative estimates of the depth at magnetic targets under the sediments were made from the magnetic field inclination derivative (TDR-DT250rp). Subtle local anomalies (maximums) of the DTrp field allowed the mapping of possible alterations by ferric oxides, with gas–oil interest.

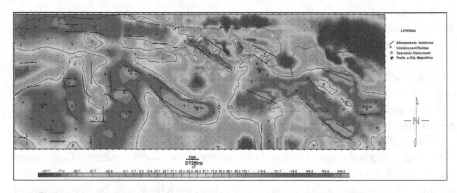

Fig. 2.9 Geological–structural mapping from magnetic data. In black thin line, limit of regionalization; in black thick line, tectonic alignments by aeromagnetometry. The maximums are associated with the presence of volcanic rocks and ophiolites of the Zaza Terrain (V) and the minimums and/or the non-anomalous field, with structural depressions and carbonates of the NACM (D). Quantitative estimates of depth to magnetic targets under sediments (by TDR) are noted at various points (Color figure online)

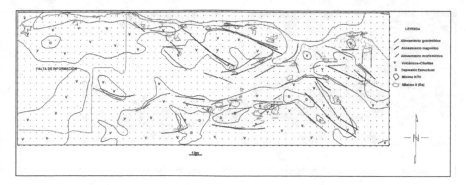

Fig. 2.10 AGS mapping, within the framework of gravimetric cartography and tectonic alignments. In red, minimum of the K/Th ratio; in pink, maximum of U(Ra); in black thin line, limit of regionalization; V, volcanic + ophiolites; D, structural depression; in gray-greenish, gravimetric alignment; in black, magnetic alignment; in blue, morphometric alignment. The green line on the map is the boundary of the non-surveyed area (Color figure online)

Airborne Gamma Spectrometry
For the AGS, the potassium (K) minima were determined, and they were compared with the minima of the K/Th ratio and the U channel (Ra), in order to follow a discriminatory purpose, indicating the localities, presumably linked, with active zones of vertical microseepage of light hydrocarbons. The results of the AGS mapping, within the framework of gravimetric mapping and tectonic alignments, are presented in Fig. 2.10.

Non-conventional Morphometry
The DEM (90 × 90 m) was subjected to a regional–residual separation from the AAC at 500 m, according to the author's experience, to plot the tectonic alignments from the total horizontal derivative DEMTHD field. The local maximums in the residual component DEMres500 field are also determined, which could be linked to the possible zones of active microseepage of light hydrocarbons previously mapped (Fig. 2.10). The regional component DEMreg500 field, its tectonic alignments and the local maxima of the DEMres500 field are presented in Fig. 2.11.

Integrated Prospective Mapping and Favorable Sectors
In the integrated prospective mapping, in order to establish the favorable sectors for the occurrence of hydrocarbons, the results of the AGS mapping were considered together with the local gravimetric, magnetic, morphometric maxima and RS anomalies (indicator attributes), in the framework of gravimetric mapping and a selection of tectonic alignments (Fig. 2.12). The favorable sectors related to oil and conventional gas of the Camajuaní and Placetas TSUs are shown (after comparing them with the geological data), also in the same framework (Fig. 2.13).

In order to offer a more realistic image of the results obtained, the favorable sectors are compared with the petroleum information of the territory (Fig. 2.14).

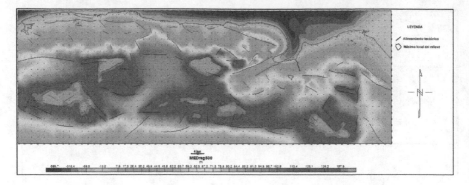

Fig. 2.11 Regional component DEMreg500 field of the study area. In blue, tectonic alignments by morphometry and contour of local morphometric maxima of possible gas–oil interest (few) (Color figure online)

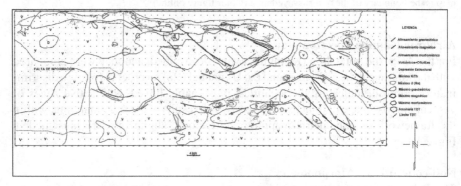

Fig. 2.12 Results of integrated prospective mapping, within the framework of gravimetric mapping and a selection of tectonic alignments. The green line on the map is the boundary of the non-surveyed area. In red, minimum of the K/Th ratio; in pink, maximum of U(Ra); in black thin line, limit of regionalization; V, volcanic + ophiolites; D, structural depression; in gray-greenish, gravimetric alignment and local maxima; in black, magnetic alignment and local maxima; in blue, morphometric alignment and local maxima; in brown, RS anomaly and limit (Color figure online)

2.4 Discussion

2.4.1 Land Exploratory Block 6

The possible structural highs determined in this research form chains or stripes of direction WSW–ENE (Elevados A and B, Fig. 2.15). The structural highs alternate with possible depressed areas, the most important being Los Palacios Basin, to the south of the Pinar Fault, southern limit of Elevado A. The thickness of the sediments in this section of the referred basin, between 1000 and 1300 m, is established by the magnetic field interpretation (TDR) in the SE zone of the CVA (Fig. 2.16).

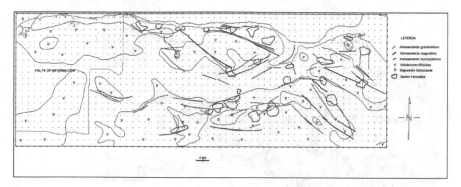

Fig. 2.13 Favorable sectors for hydrocarbons, within the framework of gravimetric mapping and a selection of tectonic alignments. The green line on the map is the boundary of the non-surveyed area. In black thin line, limit of regionalization; V, volcanic + ophiolites; D, structural depression; in gray-greenish, gravimetric alignment; in black, magnetic alignment; in blue, morphometric alignment; in purple, favorable sector (Color figure online)

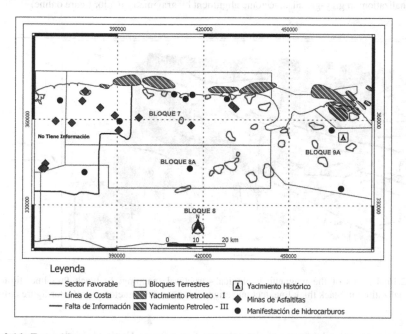

Fig. 2.14 Favorable sectors for hydrocarbons within the framework of the territory's petroleum information. The black line on the map is the boundary of the non-surveyed area. In brown, limit of the land exploration blocks; in blue, coast line; in purple, favorable sectors; red polygon, oilfield Family I; blue polygon, oilfield Family III; brown rhombus, asphaltite mine; black dot, hydrocarbon show; black rectangle, historical oilfield (Color figure online)

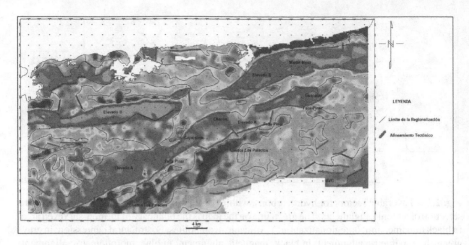

Fig. 2.15 Results of the gravimetric geological–structural mapping. In black thin line, limit of regionalization; in gray-greenish, tectonic alignment by gravimetry (Color figure online)

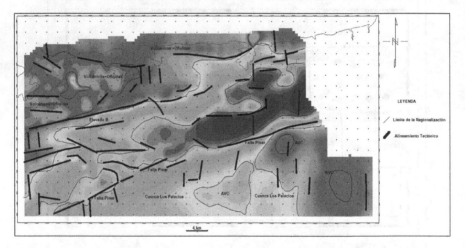

Fig. 2.16 Results of the magnetic geological–structural mapping. In black thin line, limit of regionalization; in black thick line, tectonic alignment by aeromagnetometry (Color figure online)

Judging by the location of the traces of the Pinar Fault from the gravimetric, magnetic and morphometric data, it has an S-SE dip and a normal character.

Due the wide distribution of the minimums of the K/Th ratio, the greatest gas–oil interest is proposed for Elevado A, at the WSW of the Cayajabos structure, where, also, the morphometric data (Fig. 2.6) indicates the existence of a possible less conserved block (lower erosion cut). A particular local gas–oil interest should be considered for those localities where the minimums of the K/Th ratio coincide, with maximums of U(Ra), local gravimetric and morphometric anomalies (maximums).

2.4.2 Land Exploratory Blocks 7, 8A and 9A

When considering the geological–structural position of the favorable sectors (Fig. 2.13) draws attention, first of all, a northern trend of latitudinal direction with eight sectors, linked, apparently, to the known deposits of the Cuban North Oil Belt (CNOB) and to the NACM sequences. These eight sectors have a first priority in the exploratory interest given their acceptable correspondence with the sites of oil shows, asphaltite mines and with the Family I oilfields associated with rocks of the NACM (Fig. 2.14). It should be noted that the petroleum systems that have generated Family I oil have been capable of forming giant deposits (Varadero oilfield), large and medium in the national territory (Colectivo de Autores, 2009b).

A second priority in the exploratory interest is another trend (with five favorable sectors and petroleum data), also in a latitudinal direction, at the level of Block 8A.

The interest of the remaining favorable sectors is based on their possible relationship with mapped seismic structures (not considered in this work).

Finally, the structural picture of the region, revealed by gravimagnetic and morphometric data, reveals the predominance of geological structures in the Cuban direction (NW–SE) and others of heading NE-SW (e.g., Varadero transverse fault) and latitudinal (e.g., Habana–Matanzas south depression, heading EW).

2.5 Conclusions

- The geological–structural cartography of the Block 6 region, using gravimagnetic data, establishes that the possible structural highs form chains or stripes in the WSW–ENE direction (Elevados A and B), which alternate with presumable depressed areas.
- A version of the cartography of sectors of oil and gas interest at Block 6 region, based on the presence of a complex of indicator geophysical–morphometric anomalies, reveals Elevado A, to the W-SW of the Cayajabos structure, as the main area of interest.
- A version of the cartography of favorable sectors linked to conventional oil of the Camajuaní and Placetas TSUs was obtained, based on the presence of a complex of indicator anomalies for the Habana–Matanzas region (Block 7, 8A and 9A). The applied methodology (remote sensing, gravimetry, aeromagnetometry, airborne gamma spectrometry and non-conventional morphometry) revealed two alignments of favorable sectors for oil exploration:

  – Northern alignment of latitudinal direction with eight sectors, linked to the sequences of the NACM and to the known deposits in the CNOB.
  – Southern alignment of latitudinal direction with five sectors, linked to the Zaza sequences.

- The anomalous complex considers: minimums of the K / Th ratio and local maximums of U (Ra) in a majority in its periphery; magnetic maximums (possible presence of iron oxides); morphometric maximums (possible presence of carbonatization and silicification) and; RS anomalies (possible presence of iron oxides and carbonatization), all related to possible vertical zones of active microseepage of light hydrocarbons. Weak gravimetric maximums are added to them, linked to presumable local structural uplifts. This information constitutes an essential complement for the reinterpretation of the 2D seismic works carried out recently in the territory.
- By means of the applied methodology, the geological–structural mapping of Habana–Matanzas region was achieved, based on gravimagnetic and morphometric data, which was one of the main results. From them the structural picture of this region is clarified, where the geological structures of Cuban direction (heading NW–SE) and others of heading NE-SW and latitudinal (heading E-W) predominate.

Acknowledgements We are grateful to the institution, Centro de Investigación del Petróleo, for allowing the publication of partial, non-confidential information on various research projects in the Block 6 and Habana–Matanzas (Blocks 7, 8A and 9A) regions.

Thanks for the review of the manuscript and for the observations made to it to: Dr. Osvaldo Rodríguez Morán; Dr. Reinaldo Rojas Consuegra; Dr. Evelio Linares Cala; Dr. Juan Guillermo López Rivera and M.Sc. Orelvis Delgado López.

References

Colectivo de Autores (2008) Mapa digital de las Manifestaciones de Hidrocarburos de la República de Cuba a escala 1:250,000. Unpublished report. Centro de Investigaciones del Petróleo, La Habana

Colectivo de Autores (2009a) Mapa digital de los Pozos Petroleros de la República de Cuba a escala 1:250,000. Unpublished report. Centro de Investigaciones del Petróleo, La Habana

Colectivo de Autores (2009b) Expediente Único del Proyecto 6004, "Exploración en la Franja Norte Petrolera Cubana". Unpublished report. Archivo, Centro de Investigación del Petróleo (Ceinpet), La Habana, Cuba, p 22

Colectivo de Autores (2010) Mapa Geológico Digital de la República de Cuba a escala 1:100 000. Unpublished report. Instituto de Geología y Paleontología, Servicio Geológico de Cuba, La Habana

Domínguez Gómez A, Domínguez Garcés R, Aballí Fortén P et al (2005) Informe Integral sobre la interpretación de los datos geólogo—geofísicos del Bloque 6. Unpublished report, Digicupet, La Habana, Cuba, p 36

Ducloz C (1960) Mapa Geológico de Matanzas a escala 1:20,000. Centro Nac. Fondo Geol., Minist. Indust. Bas., La Habana (unpublished report)

Goto K, Tada R, Tajika E, Iturralde Vinent MA, Matsui T, Yamamoto S, Nakano Y, Oji T, Kiyokawa S, García Delgado D, Díaz Otero C, Rojas Consuegra R (2008) Lateral lithological and compositional variations of the cretaceous/tertiary deep-sea tsunami deposit in northwestern Cuba. Cretac Res 29(2):217–236

Hatten CW, Somin ML, Millán Trujillo G, Renne P, Kistler RW, Mattinson JM (1988) Tectonostratigraphic units of central Cuba In: Barker L (ed) Transactions of the 11th Caribbean geological conference, Barbados, 1986: pp. 35.1–35

Jiménez de la Fuente L (2017) Interpretación preliminar de sensores remotos en las áreas de Guanabo-Jibacoa y Canasí-Seboruco. Reporte de Investigación. Unpublished report. Centro de Investigación del Petróleo, La Habana, p 3

Mondelo Diez F, Sánchez Cruz R et al (2011) Mapas geofísicos regionales de gravimetría, magnetometría, intensidad y espectrometría gamma de la República de Cuba, escalas 1:2,000,000 hasta 1:50,000. Unpublished report. IGP, La Habana, p 278

Pardo Echarte ME, Rodríguez Morán O (2016) Unconventional methods for oil & gas exploration in Cuba. Springer Briefs Earth Syst Sci. https://doi.org/10.1007/978-3-319-28017-2

Pardo Echarte ME, Cobiella Reguera JL (2017) Oil and gas exploration in Cuba: geological-structural cartography using potential fields and airborne gamma spectrometry. Springer Briefs Earth Syst Sci. https://doi.org/10.1007/978-3-319-56744-0

Pardo Echarte ME (2019) Reporte de investigación sobre los resultados de la aplicación de los métodos no-sísmicos de exploración en el Bloque 6, Cuba Occidental. Unpublished report. Centro de Investigación del Petróleo, La Habana, p 17

Pardo Echarte ME, Rodríguez Morán O, Delgado López O (2019) Non-seismic and non-conventional exploration methods for oil and gas in Cuba. Springer Briefs Earth Syst Sci. https://doi.org/10.1007/978-3-319-28017-2

Price LC (1985) A critical overview of and proposed working model for hydrocarbon microseepage. US Department of the Interior Geological Survey. Open-File Report 85-271

Sánchez Cruz R, Mondelo Diez F et al. (2015) Mapas Morfométricos de la República de Cuba para las escalas 1:1,000,000–1:50,000 como apoyo a la Interpretación Geofísica. Memorias VI Convención Cubana de Ciencias de la Tierra, VIII Congreso Cubano de Geofísica. Source: http://www.cgiar-csi.org/data/srtm-90m-digital-elevation

Saunders DF, Burson KR, Thompson CK (1999) Model for hydrocarbon microseepage and related near-surface alterations. AAPG Bull 83(1):170–185

Schumacher D (1996) Hydrocarbon-induced alteration of soils and sediments. In: Schumacher D, Abrams MA (eds) Hydrocarbon migration and its near-surface expression: AAPG Memoir 66, pp 71–89

Tada R, Iturralde Vinent MA, Matsui T, Tajika E, Oji T, Goto K, Nakano Y, Takayama H, Yamamoto S, Kiyokawa S, Toyoda K, García Delgado D, Díaz Otero C, Rojas Consuegra R (2003) K/T boundary deposits in the Paleo-western Caribbean basin. In: Bartolini C, Buffler RT, Blickwede J (eds) The Circum-Gulf of Mexico and the Caribbean: hydrocarbon habitats, basin formation, and plate tectonics: AAPG Memoir 79, pp 582–604

Chapter 3
Identification and Mapping of Hydrocarbon Microseepage by Remote Sensors in the Majaguillar–Motembo Region, Cuba

Lourdes Jiménez de la Fuente, Manuel Enrique Pardo Echarte, and Julio Gómez Herrera

Abstract The remote sensors allow the analysis of the terrain in order to identify indications of possible hydrocarbon microseepage in soils and sediments. Microseepage are invisibly hydrocarbon escapes that are manifested on the surface through changes in the reflectance, stress on vegetation, abnormal concentrations of kaolinite, iron oxides and carbonate alterations. The objective of the work is to identify areas of possible hydrocarbon microseepage from the study of the last four indirect indices through the analysis of optical images. The digital processing of multispectral images, Aster, Landsat 7 and 8 consisted of red green blue (RGB) combinations, band ratios and integration and analysis of the information in the geographic information system. Spatial–temporal studies of normalized difference vegetation index (NDVI), analysis of thermal reflectivity images of the surface related to the stress of the vegetation and studies of mineralogical anomalies were carried out at a scale of 1:50,000. Areas with microseepage near the Motembo oilfield and the zone of Menéndez in Villa Clara province were interpreted, being corroborated by data from surface geochemical surveys with the presence of gas in soils. Another area with microseepage was identified in the Majaguillar area, with similar patterns to the previous ones, which was supported by the interpretation of a complex of non-seismic and non-conventional methods. In addition to these, other areas were interpreted with a lower degree of confidence but with characteristics very similar to those already established.

L. Jiménez de la Fuente (✉) · M. E. Pardo Echarte · J. Gómez Herrera
Centro de Investigación del Petróleo, Churruca, No. 481, e/ Vía Blanca y Washington, 12000 El Cerro, La Habana, CP, Cuba
e-mail: lourdes@ceinpet.cupet.cu

M. E. Pardo Echarte
e-mail: pardo@ceinpet.cupet.cu

J. Gómez Herrera
e-mail: juliog@ceinpet.cupet.cu

Keywords Remote sensing · Hydrocarbon microseepage · Geographic information system · Normalized difference vegetation index · Mineralogical anomalies

Abbreviations

RGB	Red green blue
NDVI	Normalized difference vegetation index
CEINPET	Centro de Investigación del Petróleo
IDP	Image digital processing
FLAASH	Fast line of sight atmospheric analysis of hypercubes
GIS	Geographic information system
NIR	Near infrared
Red	Visible red
LST	Land surface temperature
VIS	Visible region of the electromagnetic spectrum
SWIR	Short wave infrared
TIR	Thermic infrared
TSU	Tectonic stratigraphic unit
OLI/TIRS	Operational land imager/thermal infrared sensor
ETM	Enhanced thematic mapper

3.1 Introduction

Hydrocarbon leaks (gas and/or oil) are of great importance for exploration as they are indicators of accumulations in the subsoil; their study allows to determine the quality of the oil before drilling the wells. The vast majority of oil and gas accumulations are dynamic and their seals are imperfect, producing migrations to the surface. The surface shows of hydrocarbons can be active or passive, visible (macro) or only analytically detectable (microleaks). Gases can move vertically through strata of meters in a relatively short time (weeks or years) (Schumacher 2008). The presence of surface microseepage is a direct indicator of the migration process and its study provides important indications of the active petroleum systems present and the nature of the microseepage (Salati 2014).

The first antecedents of the application in Cuba of remote sensors for the identification and mapping of hydrocarbon microseepage are in the example of non-seismic exploration in the Habana–Matanzas region (Pardo Echarte et al. 2019).

The objective of this work is to identify and map hydrocarbon microseepage from a multisensory optical imaging study in the Majaguillar–Motembo region, Cuba.

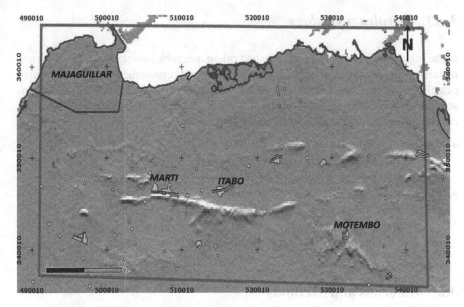

Fig. 3.1 Location of the Majaguillar–Motembo study region

3.1.1 Location of the Study Region

The studied area comprises the Majaguillar–Motembo region located in the Matanzas and Villa Clara provinces and is located in the Lambert coordinates of the Cuba North projection (Fig. 3.1).

X: 489,728; Y: 362,892.
X: 567,675; Y: 334,702.

3.1.2 Regional Geology and Tectonics

The Majaguillar–Motembo region is characterized by the development of formations whose ages range from the Upper Jurassic to the Miocene, represented mainly by carbonate sequences, policomponent olistostromes and sandy sequences. Its geological structure is represented by thrusted mantles and the superposition of different tectonic events with their implications in the subsequent plicative and disjunctive deformations that exacerbate the structural complexity of the area. Stacked structures are developed similar to those developed in the eastern portion of the Habana–Matanzas region, which occur in synclines or frontal basins. There is also the presence of ramp anticlines originated as a result of back thrust, that is, efforts contrary to those that originated the main stacked area (Domínguez Sardiñas et al. 2016).

3.1.3 Petroleum Geology

In the region, according to the established oil-source rock correlations, the existence of three different types of mature source rocks and, consequently, of three active petroleum systems, associated with families I and III of Cuban crude oil, is demonstrated with proven category (Delgado López et al. 2011). Two of these systems are related to Family I and another to Family III:

- Related to Family I:
    1. Grupo Veloz oil system—Grupo Veloz (!), linked to Placetas TSU.
    2. Jaguita/Margarita—Sagua oil system (!), linked to Camajuaní TSU.
- Related to Family III:
    1. Carmita oil system—Ophiolites (!), Linked to Placetas TSU.
    Of these systems, the most important is Grupo Veloz—Grupo Veloz (!) as the one with the highest load capacity demonstrated in the Northern Cuban Oil Belt.

3.2 Materials and Methods

3.2.1 Materials

The following images were used:

- From the Landsat 7 sensor:
    LE07_L1TP_p015r044_20010311_20040212_01.
    LE07_L1TP_015044_20140126_20161118_01.
    LE07_L1TP_015044_20200127_20200222_01.
- From the Landsat 8 sensor:
    LC08_L1TP_015044_20140118_20170306_01.
- From the Sentinel 2 sensor:
    S2B_MSIL2A_20200205T160509_N0214_R054_T17QMF_20200205T19374?
- From the ASTER sensor:
    AST_L1B_00301252002161045_20081014200655_7402.

In addition, the following were used: the generalized petroleum geology map at a scale of 1:250,000 (Colectivo de Autores 2000); the results of geophysical studies (seismic (Echevarría Rodríguez et al. 2005], non-seismic and non-conventional (Pardo Echarte and Cobiella Reguera 2017; Pardo Echarte et al. 2019)); the results of the geochemical survey carried out by Sherritt International in 1998 (in: Domínguez Sardiñas et al. 2016) and petroleum data from the Oil and Geosite Database of the Centro de Investigación del Petróleo (CEINPET) (Colectivo de Autores 1998).

3.2.2 Processing Methods and Techniques

3.2.2.1 Indirect Indices of the Presence of Hydrocarbons

Hydrocarbon migrations are defined as active and passive (Abrams 1996, in: Salati 2014). The former occur fundamentally in oil basins where there is active generation of hydrocarbons and are also called macroseepage. Macroseepage areas indicate the presence of excellent migration pathways. Passive migrations (microseepage), meanwhile, represent remnants of seepage in basins with passive generation of hydrocarbons and/or the presence of excellent seals or poor migration routes. In general, microseepage constitutes vertical or almost vertical leaks from reservoirs to the surface (Salati 2014). The vertical movement of light hydrocarbons from the reservoir rock to the surface can be through networks of fractures, faults and bedding planes that provide permeable paths within the overlying rock (Khan 2006).

The methodology for the identification and mapping of areas with microseepage in the Majaguillar–Motembo region consisted of obtaining indirect indices of the presence of hydrocarbons in soils and sediments through their response in the electromagnetic spectrum captured by the sensors used. These results were supported by geophysical and geochemical studies in the interpreted areas.

Hydrocarbons can be expressed in different ways on the surface; they can be:

- Surfaces with abnormal concentrations of hydrocarbons in sediments, soils and waters.
- Mineralogical changes such as calcite formation, clay alterations, alterations associated with iron oxides and sulfated minerals and bleaching.
- Geobotanical anomalies.

Next, the geochemical anomalies related to the presence of hydrocarbon microseepage are described.

Carbonates
Diagenetic carbonates and carbonate cementation are the most common alterations associated with hydrocarbon microfiltration in terrestrial environments. These disturbances are mainly formed by oxidation of oil, particularly methane, under aerobic or anaerobic conditions as described below (Schumacher 1996):

Aerobic:

$$CH_4 + 2O_2 + Ca_2^+ = CaCO_3 + H_2O + 2H^+$$

Anaerobic:

$$CH_4 + SO_4H_2^- + Ca_2^+ = CaCO_3 + H_2S + H_2O$$

Sulfated Minerals

Pyrite can be precipitated in reductive environments, either from the degradation of oil as a result of the action of bacteria or from the oxidation of oil near the surface. The development of the pyrite disturbance zones depends on the sulfur content in the oil, the geology and the chemical composition of the groundwater, as well as the degradation by bacteria. The reaction of hydrogen sulfide with iron (in hematite) precipitates pyrite and its reaction is summarized below (Oehlerand Sternberg 1984; Hughes et al. 1986, in Schumacher 1996):

$$Fe_2O_3 + 4H_2S = 2FeS_2 + 3H_2O + 2H^+ + 2e^-$$

Bleaching

This phenomenon promotes the discoloration of limonitic materials (rocks and soils), due to the acid-reducing action of the solutions that facilitate the elimination of the ferric ions (3^+) (ex., hematite) found in these materials. In many cases, ferric ions can be converted to ferrous ions (2^+) and favor the formation of pyrite, siderite and, finally, jarosite (Schumacher 1996).

Clay Minerals

The production of carbon dioxide, hydrogen sulfide and organic acids, which result from the microbial oxidation of hydrocarbons near the surface, generates an acid-reducing action capable of promoting the diagenetic alteration of feldspars to clay minerals, including kaolinite, illite and chlorite (Schumacher 1996). The spectral characteristic of clay minerals can also be detected in soils, as a function of the secondary effects caused by bleaching. As the high iron content is removed from the soil, the absorption characteristics of the clays tend to become more apparent.

Geobotanical Anomalies

The hydrocarbon microseepage create a reduction zone in the soil at a shallow depth that stimulates the action of bacteria, resulting in a decrease in the concentration of oxygen and an increase in the concentration of CO_2 and organic acids in the soil. These changes produce a variation in the pH and Eh of the soil and therefore in the vigor of the vegetation (Schumacher 1996).

Such alteration is commonly marked by one or more of the following characteristics:

- The lack of vegetation;
- Variations in the density of the flora;
- Variations in the architecture of the cup and
- Morphological changes in the species. (Van der Meer et al. 2002; Souza Filho et al. 2008; Sánchez et al. 2013, all of them cited in Plata 2015).

3.2.2.2   Image Digital Processing (IDP)

The preprocessing of the satellite images consisted of the following steps:

- Radiance calibration
- Layer stacking
- Subset
- Reprojection
- Reflectance calibration with the Fast Line of sight Atmospheric Analysis of Hypercubes (FLAASH) module and the Log Residuals algorithm.

All preprocessing and interpretation were performed with ENVI 4.7 and QGis 3.18 software.

3.2.2.3   Analysis of Satellite Information

For the analysis and extraction of satellite information, an index and band ratios were built, which are explained below:

Normalized Difference Vegetation Index (NDVI)
Its formulation is based on the characteristic reflectance of chlorophyll in the red and near infrared regions. The infrared range varies between -1 and 1, with the values closest to 1 being those associated with more vigorous vegetation.

$$NDVI = (NIR - Red)/(NIR + Red),$$

NIR: Near infrared.
Red: Visible red.
To define the intervals, a temporal evolution of the NDVI was taken into account, in a period from 2001 to 2020. For this work, the following ranges were assumed:

- -1.0 Dead plant or inanimate object
- 0.33 Diseased plant
- 0.33–0.66 Moderately healthy plant
- 0.66–1 Very healthy plant.

Surface Temperature
The surface temperature is inversely related to plant vigor, since evapotranspiration releases heat and reduces the temperature of the plant mass compared to the bare soil. The evolution of the surface temperature allows estimating the vigor or stress conditions of the vegetation (Chuvhieco 1996). In this study, Landsat 7 and 8 images were used to calculate the surface temperature. This parameter, as well as the vegetation, was taken for the period 2001–2020.

Vegetation Stress

It is determined from the behavior of the NDVI and the temperature of the land's surface (LST). Those areas where the surface temperature has had an increase in the period of time 2001–2020 and where losses or absence of vegetation are defined in the same interval are delimited.

Iron Oxides and Sulfated Minerals (Guo and Mason 2005)

Iron Oxide Index: Red − min(Red)/Blue − min(Blue) + 1.

This phenomenon can be spectrally detected by virtue of the change in the rock/soil; as a function of the loss of ferric ions, this is detectable within the ranges of the visible electromagnetic spectrum (VIS) and the near infrared (NIR). Hematite, goethite and limonite produce spectral responses with a steep slope in the VIS region. Once these minerals tend to be eliminated, a sharp decrease in the VIS reflectance gradient is observed, which can be detected using multispectral data and also the change in the rock/soil spectral behavior as a function of the substitution of ferric minerals (goethite, hematite) by ferrous minerals (siderite) (Plata 2015).

Clay Minerals

Kaolinite is an alteration mineral associated with hydrocarbon microseepage. It can be spectrally characterized by a double absorption diagnosis in the Short Wave Infrared (SWIR), one centered from 1.40–1.42 microns to 2.162–2.206 microns, in addition to another characteristic of absorption at 2.312–2.350 microns, and 2.380 microns (Plata 2015).

Carbonates

Carbonates can be remotely detected by their spectral characteristics in the SWIR or thermal infrared (TIR) range of the electromagnetic spectrum. In the SWIR range, carbonates can be characterized mainly by the absorption function at 2.34 microns and by another around 1.87, 2.0 and 2.5 microns. In the TIR interval, in particular, in the atmospheric window between 8 and 14 microns, around 11.2 microns (Salisbury et al. 1991, in: Plata 2015).

Table 3.1 summarizes the band ratios calculated for the analysis of the geobotanical and mineralogical indicators.

3.2.2.4   Processing in GIS

The data from the remote sensors were processed in QGis 3.18, to identify and map the favorable areas for the occurrence of microseepage, taking advantage of the advantages offered by GIS. In a later step, all the geological-oil information was integrated to rule out false positives in the interpretation.

Table 3.1 Band ratios used according to types of sensors

Sensors	Band ratios
Aster	4/6 clays
Aster	4/9 carbonates
Sentinel 2	NDVI vegetation
Landsat 8 OLI/ TIRS	NDVI vegetación
Landsat ETM 7	Iron oxide index
Landsat ETM 7	Surface temperature
Landsat 8 OLI/ TIRS	Surface temperature

3.3 Results

3.3.1 Analysis of Satellite Data

The vegetation stress analysis illustrates areas with diseased vegetation and areas with loss of cover in the period 2001–2020. Regarding its distribution, a higher density of vegetative stress is located toward the NW part of the region; however, this condition is mainly due to the increase in agricultural areas.

From the analysis of the remote sensors, a set of mineralogical anomalies indicative of the presence of microseepage was determined. A well-defined area stands out toward the Majaguillar sector where alterations by carbonates and clays mainly combine. Another outstanding element is the accumulation of carbonates parallel to the coast, which must be the result of the carbonate geological formations present there. Toward the NW area, there is a greater number of anomalies due to iron or sulfate alterations (Fig. 3.2).

All these elements made it possible to identify satellite anomalies associated, presumably, to the presence of hydrocarbon microseepage. A total of 80 satellite anomalies in the Majaguillar–Motembo region were defined by remote sensing analysis and integration with GIS. These were mapped from the spatial coincidence of different indicators of the presence of hydrocarbons.

3.4 Discussion

3.4.1 Interpreted Sectors

A total of four favorable sectors for the occurrence of hydrocarbon microseepage were defined, to which orders were assigned according to their level of reliability based on integration with geological-oil elements. In this sense, the qualitative classification responds to a decreasing order of reliability, being those of the first order those where

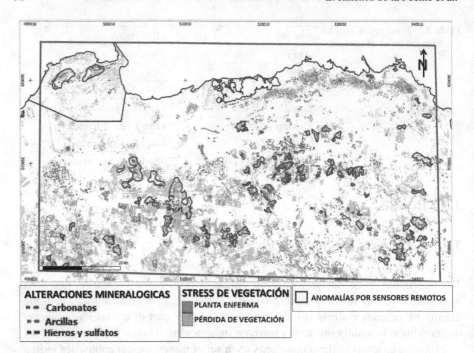

ALTERACIONES MINERALOGICAS | STRESS DE VEGETACIÓN | ANOMALÍAS POR SENSORES REMOTOS
-- Carbonatos | PLANTA ENFERMA |
-- Arcillas | PÉRDIDA DE VEGETACIÓN |
-- Hierros y sulfatos | |

Fig. 3.2 Results of satellite image processing. Anomalies interpreted from remote sensors. In blue dashed line, carbonate alterations; in a dashed brown line, clay alterations; in black dotted line, changes by iron and sulfates; in pink polygon, remote sensing anomalies; in orange polygon, diseased plants; in green polygon, areas with loss of vegetation (Color figure online)

almost all the elements of satellite-oil information are concentrated and, those of the third order, in general, present only satellite patterns with insufficient information to corroborate microseepage.

Two sectors of greatest interest were identified: Majaguillar and Motembo. In the first, Neo-autochthonous rocks emerge, characterized mainly by marsh deposits. There, the Majaguillar oilfield of conventional oil is located, related to the Placetas TSU and, above this, the Peñón sequences, linked to the presence of unconventional oil. Several oil wells confirm the potential of the sector. In the same, studies of non-seismic and unconventional methods were carried out (Pardo Echarte and Cobiella Reguera 2017) with favorable results for the occurrence of hydrocarbon microseepage and accumulations in the depth. All this set of methods and criteria of potentiality in the sector allow it to be classified as first order (Fig. 3.3).

In the second sector, Motembo (Fig. 3.4), the outcrops of the Placetas TSU, the Ophiolitic Association and the postorogenic rocks predominate. The main exploration works carried out in this area are the gaseous geochemical survey carried out by Sherritt International 1998 (in: Domínguez Sardiñas 2016) and the non-seismic and non-conventional methods (Pardo Echarte et al. 2019) carried out in the Motembo

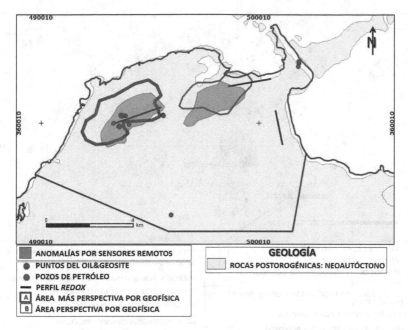

Fig. 3.3 Majaguillar sector. Results of the remote sensors and of the set of non-seismic and non-conventional methods (Pardo Echarte and Cobiella Reguera 2017). In pink polygon, remote sensing anomalies; in blue points, points from the OIL and GEOSITE database; in red dots, oil wells; in black line, profile of the Redox Complex; in green polygon, perspective area by geophysics; in yellow polygon, postorogenic rocks (Color figure online)

SE and Motembo Carbonates areas. As a result of the second area, large gaseous concentrations are present in the samples taken around the areas, where a set of satellite anomalies are identified. All these potentiality criteria allow classifying this sector as first order.

The third favorable sector, corresponding to a second order of reliability, is the Martí sector (Fig. 3.5). Geologically, the sequences of the Placetas TSU and their overlapping basins predominate, as well as postorogenic deposits, in addition to small patches where the rocks of the Ophiolitic Association emerge. It is characterized by the presence of oil shows in wells at shallow depths such as the cases of Martí 1, 2 and 5 and further south, Arabia 1. The set of satellite anomalies is spatially distributed over the deposits of the Peñón Formation, recognized due to its potential for unconventional oil. Although the data are favorable to the presence of oil and gas in soils, there is no sample, geochemical data or profile of the Redox Complex (non-conventional method) that demonstrates the presence of microseepage, likewise, there are no discoveries of oilfields in the sector.

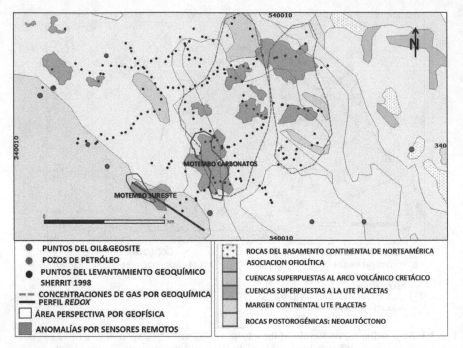

Fig. 3.4 Motembo sector. Set of non-seismic methods and Redox Complex. In pink polygon, remote sensing anomalies; in blue points, points from the OIL and GEOSITE database; in red dots, oil wells; in black points, points of the geochemical survey; in blue dotted line, gas concentrations; in black line, profile of the Redox Complex; in green polygon, perspective area by geophysics; in white and red polygon, rocks of the North American continental basement; in light purple polygon, rocks of the Ophiolitic Association; in a light green polygon, basins superimposed on the volcanic arc; in the ocher polygon, basins superimposed on the Placetas TSU; in blue polygon, limit of the Placetas TSU; in yellow polygon, postorogenic rocks (Color figure online)

The fourth favorable sector is Menéndez-Bolaños, where postorogenic deposits and the sequences of the Camajuaní and Placetas TSUs predominate. A seismic survey was carried out in the sector (Echevarría Rodríguez et al. 2005), resulting in the mapping of some positive structures. In it, two areas are differentiated, one to the east and the other to the south. In the first, a set of oil wells with the presence of hydrocarbons are located, and a gaseous geochemical survey was carried out (Sherritt International 1998, in: Domínguez Sardiñas et al. 2016). Although these data are not conclusive to demonstrate the presence of microseepage in the area, it is possible to associate, in a preliminary way, the satellite anomalies interpreted with faults, as possible routes of migration of hydrocarbons. Due to this, it was classified as second order. In the one corresponding to the south of the sector, the level of reliability decreases to a third order, since despite the fact that similar disjunctive elements are recognized, no oil wells are located in a close radius that corroborate the presence of hydrocarbon emanations (Fig. 3.6).

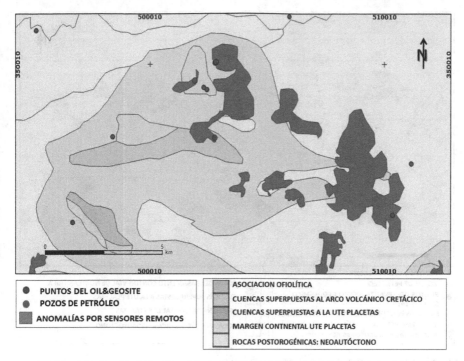

Fig. 3.5 Martí sector. The presence of hydrocarbon manifestations in oil wells stands out. In pink polygon, remote sensing anomalies; in blue points, points from the OIL and GEOSITE database; in red dots, oil wells; in light purple polygon, rocks of the Ophiolitic Association; in a light green polygon, basins superimposed on the Volcanic Arc; in the ocher polygon, basins superimposed on the Placetas TSU; in blue polygon, limit of the Placetas TSU; in yellow polygon, postorogenic rocks (Color figure online)

Finally, all the favorable sectors with the highest probability of occurrence of hydrocarbon microseepage and their confidence levels are presented, based on the analysis of mineralogical and geobotanical alterations, defined by the analysis of remote sensors (Fig. 3.7).

3.5 Conclusions

- The interpretation of optical satellite images allowed the identification and mapping of indirect indices related to the presence of hydrocarbons in soils and sediments such as vegetation stress and anomalous concentrations of clays and carbonates. A total of 80 anomalies associated with the presence of possible hydrocarbon microseepage could be identified and mapped, which were classified based on their relationship with geological-oil elements, leaving only four sectors as potential for their occurrence.

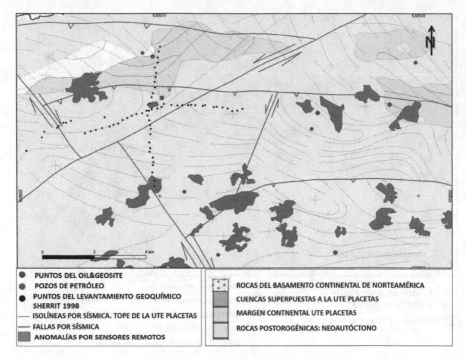

Fig. 3.6 Menéndez-Bolaños sector. The relationship of structural elements with the spatial distribution of the anomalies interpreted by remote sensors is highlighted. In pink polygon, remote sensing anomalies; in blue points, points from the OIL and GEOSITE database; in red dots, oil wells; in black points, points of the geochemical survey; in blue line, contours by the top of the Placetas TSU; in red line, faults interpreted by seismic; in white and red polygon, rocks of the North American continental basement; in the ocher polygon, basins superimposed on the Placetas TSU; in blue polygon, limit of the Placetas TSU; in yellow polygon, postorogenic rocks (Color figure online)

- The Majaguillar and Motembo sectors are related to known oil fields and with favorable results from the set of non-seismic and non-conventional exploration methods. The Martí and Menéndez-Bolaños sectors were classified as second order and are mainly related to oil demonstrations in wells. In this last sector, a less favorable area was detected, related only to structural elements, being considered of third-order.
- The identification of satellite patterns in validated sectors (Majaguillar and Motembo) allows the extrapolation of this information to others where no information is available.

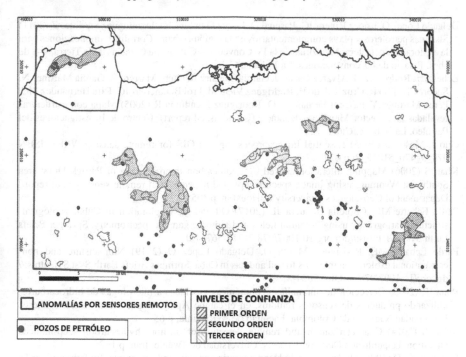

Fig. 3.7 Confidence levels of the sectors for the occurrence of hydrocarbon microseepage. In red polygon, first-order sectors; in green polygon, second-order sectors; in yellow polygon, third-order sectors; in red dots, oil wells (Color figure online)

Acknowledgements The authors thank the Centro de Investigación Del Petróleo and its Technical Archive for allowing the publication of non-confidential information on their research, as well as to Dr. Reinaldo Rojas Consuegra, and to Dr. Olga Castro Castiñeira, researchers at this institution, for the exhaustive and rigorous review of the manuscript.

References

Colectivo de Autores (1998) Base de Datos Petrolera (Oil & Geosite). (Unpublished report). Centro de Investigaciones del Petróleo, La Habana, Cuba

Colectivo de Autores (2000) Mapa generalizado de geología petrolera a escala 1:250,000 de la República de Cuba. (Unpublished report). Centro de Investigaciones del Petróleo, La Habana, Cuba

Chuvhieco E (1996) Fundamentos de teledetección espacial, ISBN 84_321_3127X. Ediciones RIALP, 3ra edición. Madrid, España, p 568

Domínguez Sardiñas Z, Linares Valdés L, Jiménez de la Fuente L, Rifa Hernández MC, Pról Betancourt JG, Pardo Echarte ME, Rodríguez Morán O, Díaz Díaz M, Rivas L, Rey Pallí R (2016) Estudios y evaluaciones para la exploración de gas no convencional en Motembo, Antón Díaz, Jarahueca y noroeste de Holguín. (Unpublished report). Centro de Investigación del Petróleo, La Habana, Cuba

Delgado López O, López Rivera JG, Pascual O, López Quintero JO, Domínguez Sardiñas Z (2011) Sistemas petroleros y plays complementarios en la región Habana-Corralillo. Implicaciones para la exploración petrolera. Memorias de la IV Convención Cubana de Ciencias de la Tierra, 4–8 de abril. Palacio de las convenciones, La Habana, Cuba

Echevarría Rodriguez G, Alvarez Castro J, Garcia Sánchez R, Otero Marrero R, Garcia Martínez N, Socorro Trujillo R, Cruz Toledo R, Rodriguez Viera M, Pról Betancourt JG, Rifa Hernández MC, Pérez Martínez Y, Pascual Fernández O, Rodriguez Menduiña R (2005) Mapa con estructuras reveladas en el sector Menéndez-Bolaños. (Unpublished report). Centro de Investigaciones del Petróleo, La Habana, Cuba

Guo J, Mason P (2005) Essential image processing and GIS for remote sensing. Wiley. ISBN 978_0_470_51032_2, p 437

Khan S (2006) Mapping alteration caused by hydrocarbon microseepage in Patrick Draw area Southwest Wyoming using image spectroscopy and hyperspectral remote sensing final report. Department of Geosciences University of Houston, p 105

Pardo Echarte ME, Cobiella Reguera JL (2017) Oil and gas exploration in Cuba: geological-structural cartography using potential fields and airborne gamma spectrometry. Springer Briefs Earth Syst Sci. https://doi.org/10.1007/978-3-319-56744-0

Pardo Echarte ME, Rodríguez Morán O, Delgado López O (2019) Non-seismic and non-conventional exploration methods for oil and gas in Cuba. Springer Briefs Earth Syst Sci. https://doi.org/10.1007/978-3-030-15824-8

Plata IR (2015) Detección de anomalías geobotánicas asociadas a microfugas de hidrocarburos, utilizando productos de sensores remotos, en el campo Apiay. Unpublished Master's Thesis. Universidad Nacional de Colombia, Facultad de Agronomía, p 65

Salati S (2014) Characterization and remote detection of onshore hydrocarbon seep-induced alteration. Unpublished Docotoral Thesis. Universidad de Twente, Iran, p 162

Schumacher D (1996) Hydrocarbon-induced alteration of soils and sediments. In: Schumacher D, Abrams MA (eds) Hydrocarbon migration and its near surface expression, AAPG Memoir 66, pp 71–89

Schumacher D (2008) Non-seismic detection of hydrocarbons: an overview. In: AAPG search and discovery. Cape Town, South Africa, Article #40392

Chapter 4
Physical-Geological Modeling of Potential Fields in the Northeastern Region of the Central Basin of Cuba

Jessica Morales González, Manuel Enrique Pardo Echarte, and Osvaldo Rodríguez Morán

Abstract The Central Basin of Cuba was the largest oil-producing region in the country during the 1960s. However, after the 1990s with the discovery of the Pina oilfield, there has been no other significant discovery. Exploration failures are considered to be conditioned, in part, by the high geological complexity of the region and by the volcanic nature of the sequences present, which limit the depth of reflection seismic research. Thus, the problem lies in the need to use 2D physical-geological modeling of potential fields in order to help clarify the deep constitution of the territory, given the interest for finding conventional oil and gas from the Tectonostratigraphic Units (TSUs) of Camajuaní and Placetas at the sediments of the North American Continental Margin. Geological and petrophysical data, seismic data and potential fields of the northeastern region of the Central Basin of Cuba were evaluated in the preparation and interpretation of three 2D models of potential fields: one that is longitudinal to the basin and two that cut the Cristales and Pina oilfields, respectively. As a result, from the 2D physical-geological models, the hypothesis of the existence in the whole basin of carbonates from the North American Continental Margin, considered as source rock, is validated. According to the models, the top of these rocks is located, at the Pina sector, between 2.98 and 4.3 km, while at the Jatibonico-Cristales and Catalina sectors, they range between 5.55 and 6.6 km and 6.2 km, respectively. In addition, their thickness decreases from north (5 km) to south (1.3 km) and, conversely, the one of Zaza Terrain. This reinforces the hypothesis of the best prospects for finding conventional oil from the Camajuaní and Placetas TSUs in the Pina sector.

Keywords Gravimetry · Magnetometry · 2D physical-geological models of potential fields · Volcanic rocks

J. Morales González (✉) · O. Rodríguez Morán
Universidad Tecnológica de La Habana "José Antonio Echeverría" (Cujae) Calle, 114 no. 11901 entre Ciclo Vía y Rotonda, 11500 MarianaoLa Habana, CP, Cuba

M. E. Pardo Echarte
Centro de Investigación del Petróleo, Churruca, No. 481, Vía Blanca y Washington, 12000 El Cerro, La Habana, CP, Cuba
e-mail: pardo@ceinpet.cupet.cu

© The Author(s), under exclusive license to Springer Nature Switzerland AG 2022   63
M. E. Pardo Echarte et al., *Geological-Structural Mapping and Favorable Sectors for Oil and Gas in Cuba*, SpringerBriefs in Earth System Sciences,
https://doi.org/10.1007/978-3-030-92975-6_4

Abbreviations

NACM	North American Continental Margin
CVA	Cretaceous volcanic arc
OA	Ophiolite association
TSU	Tectonic stratigraphic unit
Gb	Bouguer gravity field
GbTHD	Total horizontal derivative of Gb
GbVD	First-order vertical derivative of Gb
AAC	Ascendent analytical continuation
Gbres12000	Residual Gb field from AAC at 12,000 m
DTrp (RP)	Reduced to the pole magnetic field
DTrpVD	First-order vertical derivative of DTrp
DTrpres12000	Residual DTrp field from AAC at 12,000 m
GPC	Gabbro-plagiogranite complex
CEINPET	Centro de Investigación del Petróleo

4.1 Introduction

The first successful investigations for hydrocarbons in the Central Basin, according to Martínez Rojas et al. (2006), date from the 1950s when North American companies discovered the Jatibonico (1954), Cristales (1955) and Catalina (1956) oilfields. In the 1960s, the Central Basin was the largest oil-producing region in the country, which led to an accelerated development of exploration works, which have spanned more than 50 years of research and development. However, after the 90s with the discovery of the Pina oilfield, there has been no other important finding, in part, it is due to the high geological complexity of the territory and to the volcanic character of the present sequences, which limits the depth of research of reflection seismic.

Yanquiel Martínez (2005) and Aliuska Peña (2005) carried out a 3D inversion of gravimetric data from the southern and northern part of the Central Basin. The first zone corroborates aspects related to the depth of sedimentary rocks, their thickness variations and the Bahamas Paleomargin or North American Continental Margin (NACM). The fact that in all sectors, the cross sections in the W-E direction better reflect geological and geometric heterogeneities is due to the location characteristics of the rocks of the Cretaceous Volcanic Island Arch (CVA) and the Ophiolitic Association (OA). According to the sections of the models, a homogeneous advance of the thrust front of these rocks did not occur on carbonates of the NACM, which occurred in a southwest to northeast direction. The process was also characterized by differentiated efforts. In the second zone, aspects related to the location direction of the CVA rocks are suggested, plus the OA that rides on the terrigenous-carbonate and carbonate sequences of the NACM, with a southwest to northeast direction. The deep existence of normal faults in the form of blocks with a NE–SW direction was confirmed, limited to the northeast by the Central Basin, which explains the structural development of the semigraben formed during the evolution of the basin in the Cenozoic.

In 2007, Aliuska Peña et al. (2007) performed the 3D inversion of the entire basin, determining that the depths and thicknesses of the sedimentary rocks increase toward the southeast, grow less toward the northwest part and they locate their smallest thicknesses and depths in the upper part of the folded sequences of the northwest region. Hence, the behavior of negative intensities of the gravimetric field in the southeastern sector, which is due to the existence of large thicknesses of sparse rocks above the volcanogenic and terrigenous-carbonate sequences of the NACM.

Israel Cruz Orosa (2012), through the results obtained in his doctoral research "Synorogenic basins as a record of the evolution of the Cuban Orogen: Implications for hydrocarbon exploration," developed several 2D physical-geological models of potential fields. The 2D inversion of gravimetric data constrained with surface structural data, soundings and some seismic sections allowed the author to make a proposal for a structural model, as well as the evolution of the La Trocha fault zone and define some of the characteristics of the fault zone, such as its geometry, the style of deformation and structural evolution.

José A. Batista Rodríguez et al. developed in (2014), the 3D inversion of gravimetric data of the Central Basin where it defines structural highs which could function as hydrocarbon traps of interest for a later exploration.

The problem posed to this research lies in the need to use 2D physical-geological modeling of potential fields in order to help clarify the deep constitution of the territory, with its consequent gas-oil implications. In this way, the geological and petrophysical data, the seismic and the potential fields of the northeastern region of the Central Basin of Cuba are evaluated, in order to integrate them in the preparation and interpretation of three 2D physical-geological models of potential fields: one that is longitudinal to the basin and two that cut to the Cristales and Pina oilfields, respectively.

The article begins with an introduction, in which the antecedents in the investigations carried out in the territory are exposed, then it goes on to the description of the geological framework, the materials and methods used, and the results and the discussion of these are presented, where 2D physical-geological models of potential fields are interpreted.

4.1.1 Geological Framework

From the geological point of view, in the Central Basin, the development of siliciclastic and pyroclastic rocks from the Eocene and Upper Cretaceous, respectively, is observed in the southern half and, in the northern half, the development of sediments from the Miocene and Quaternary, which they cover the oldest rocks (García and Valdés 2004). The geological constitution is supported by the drilling of several wells and by surface data, recognizing the following petrotectonic complexes:

- Postorogenic sediments (Middle Eocene to Recent).
- Non-orogenic sediments belonging to basins of the Piggy Back type (Upper Cretaceous—Paleogene).
- Zaza Terrain (CVA and OA) rocks.
- NACM rocks.

In Central Cuba, according to Cruz Orosa (2012), the synorogenic basins were formed in a context of collision. From a tectonostratigraphic analysis, different structural domains present in the Central Cuban Gold Belt were identified. These structural domains are: the Escambray Metamorphic Complex, the Axial Zone, and the North Deformation Belt (Cuban North Thrusted and Folded Belt).

The northern domain or North Deformation Belt was formed in a compressive tectonic regime. It consists of a thin-skinned interwoven system involving NACM sequences and occasionally some elements of the Caribeana and Zaza terrains, forming from the Paleocene to the Middle Eocene. The folds and faults occurred in a normal sequence, with the tectonic transport directed toward the NNE. Some SW–NE structures are contemporaneous with the imbricated system that extends in a NW–SE direction, forming tectonic corridors and/or transfer faults that facilitated a deformation partition regime while the collision occurred (Cruz Orosa 2012).

The NACM sequences were embedded in a foreland thrust belt and in some places are partially covered by allochthonous sheets of volcanic rocks and ophiolites from the Zaza terrain. These allochthonous sheets were placed out of sequence and mark the suture area between the CVA and the NACM units (Cruz Orosa 2012).

In Central Cuba (Las Villas and Camagüey blocks), the continental margin sequences are correlated with those of the Bahamas (Hatten et al. 1958; Meyerhoff and Hatten 1974, 1968). These appear tectonically covered by the Zaza Terrain. Tectonic transport in both blocks occurred mainly toward the NNE. The accretion of tectonic plates in Las Villas is well argued while in Camagüey, there is no conclusive evidence on the occurrence of tectonic accretion. Cruz Orosa et al. (2012a) suggested that the shortening in the Las Villas deformation belt was about 25 km, not including the shortening of the Tectonostratigraphic Units (TSUs) of Remedios and Cayo Coco. The shortening in the Camagüey block has not been quantified.

On the other hand, it is considered that the origin of the tectonic corridors was a consequence of the oblique convergence that gave rise to the contemporary formation of the Cuban Orogen and the Yucatán basin. According to Cruz Orosa et al. (2012b) it is considered that the La Trocha corridor was the first to be formed. It extends through the Yucatán basin as a trans-basin fault. Rosencrantz (1990) defined this fault zone based on lineaments in the relief of the ocean floor and suggested its transcurrent nature. La Trocha corridor and its extension in the Yucatán basin differentiate two regions. In the Cuban Orogen, the structural evolution on both sides of the corridor was different, while the trans-basin fault constitutes a tectonic boundary between the western and eastern domains of the Yucatan basin. The evolution of La Trocha corridor is recorded in the syntectonic sedimentation since the Paleocene and is consistent with a NW–SE extension in the Yucatán basin.

La Trocha corridor was formed due to the fact that the extreme northwest of the Caribbean became detached from the rest of the Caribbean Plate as the obliquity of the collision with the North American plate increased, varying the orientation of the transforming boundary by more than 60° clockwise. This rotation was a result of the tectonic escape of the Caribbean Plate between the North and South American plates and played a fundamental role in the sequence of accretion of the Cuban Orogen to the North American Continental Margin. This produced a NW–SE extension that led to the initial phase of the Yucatán basin during the Paleocene (Pindell et al. 2005).

In particular, La Trocha fault system has acted as a sinestral transfer zone that separates the Las Villas and Camagüey blocks in central Cuba. The structures that make up this fault system (La Trocha, Zaza-Tuinicú, Cristales and Taguasco faults) are consistent with the clockwise rotation of convergence and shortening in central Cuba. From the Paleocene to the Lower Eocene, a shortening in the SSW–NNE direction produced transtension in the La Trocha fault and transpression in the Zaza-Tuinicú fault. Later, during the Middle Eocene, the shortening rotated in a SW–NE direction, resulting in a normal component in the La Trocha fault and transpression in the Zaza-Tuinicú and Cristales faults. Since the Upper Eocene, central Cuba has been welded to the North American Plate (Cruz Orosa et al. 2012a).

Postweld deformation has produced transtension in the La Trocha and Taguasco faults and is consistent with a shortening in the WSW–ENE direction that reflects the activity of the transforming limit of Caimán. The kinematics of the plates and the structural evolution of the La Trocha fault system indicate that the Central Basin is a polygenetic tear basin and that the formation of this system (that is, fault system—tear basin) was a consequence of the oblique collision that occurred during the Paleogene between the CVA and the NACM (Cruz Orosa 2012).

4.2 Materials and Methods

4.2.1 Materials

- Geological, geophysical, geochemical and geomorphological reports of the Central Basin, Cuba.
- Reports and thematic publications on geological and oil research in the region.
- Grids of the gravitational and magnetic fields at a scale of 1:50,000 of the territory (Mondelo Diez et al. 2011).
- Digital Geological Map of the Republic of Cuba at a scale of 1:100,000 (Colectivo de Autores 2010).
- Tectonic map of the basement of Cuba (Cobiella Reguera 2017).
- Digital maps of the hydrocarbon manifestations (Colectivo de Autores 2008) and of the oil wells of the Republic of Cuba (Colectivo de Autores 2009).

4.2.2   Methods and Techniques

The method used was 2D physical-geological modeling of potential fields.

The processing technique used was the Oasis Montaj software from GeoSoft v. 8.3 and its potential field modeling module GMSYS.

4.3   Results and Discussion

4.3.1   Response of Potential Fields in the Central Basin

The gravitational field in the Central Basin has been widely studied by different authors from surveys at different scales (Kireev 1963; Ipatenko 1968; Rodríguez and Prol Betancourt 1980; Rodríguez and Domínguez 1993). The present investigation used the Grid of the Bouguer Anomaly at a scale of 1:250,000 with topographic correction for the density of 2.3 t/m^3, made from the data of Mondelo Diez et al. (2011). The mean square error of the Bouguer anomalies was $\pm$ 0.4 mGal.

The basin is identified by a large negative anomaly, in a SW-NE direction. It is divided by a maximum associated with the structural uplift of Jatibonico, which separates it into two gravitational minima (Fig. 4.1). The minimum values of Bouguer anomaly oscillate between -32 mGal for the depocenter located to the SW and, $-$

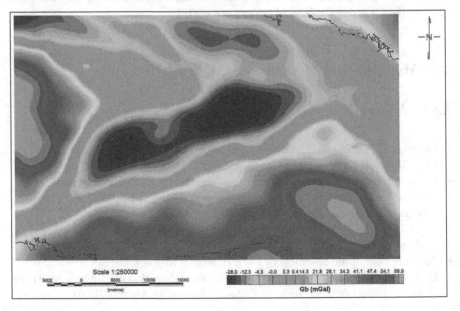

Fig. 4.1 Bouguer anomaly map ($\sigma = 2.3$ t/m^3) of the Central Basin

37 mGal for the one to the NE. Despite the fact that both anomalies are similar from the geophysical point of view, the data from wells show that they are generated by different geological behaviors, so the basin must be studied dividing it into two sectors. The Sancti Spíritus well is located in the first depocenter, which cut more than 3000 m of sediments without reaching the volcanics; while in the second depocenter, these thicknesses are lower, obtaining that for the La Rosa 3 and Pina Norte 1 wells, the sediments only have thicknesses of 1602 m and 1348 m, respectively.

It is considered that the presence of uplifts of high-density rock such as ophiolites, volcanic rocks and carbonates from the Remedios, Camajuaní and Placetas TSUs, can cause weak local gravimetric maximums. Also, it is understood that large thicknesses of syn- and postorogenic sediments caused local minima.

From the total horizontal derivative (THD) (Fig. 4.2) and the first-order vertical derivative (VD) of the gravitational field (Fig. 4.3), the main gravimetric alignments associated with faults and the limits of the main structures present are identified. Thus, the Taguasco, Zaza-Tuinicú, La Trocha and Cristales faults that constitute the edges of the Central Basin, as well as the Las Villas fault, located in the northern part, and other minor faults resulting from the activity of the La Trocha fault system were mapped.

The residual of the gravitational field at 12,000 m (Fig. 4.4) shows that the large volcanic bodies located to the west (Máximo de Fomento) extend to more than 12 km deep, as do the granitoids that emerge to the SE and the volcanic ones that are located to the south. The strong gradient changes that mark the edges of the basin in this figure coincide with the faults of the La Trocha fault system.

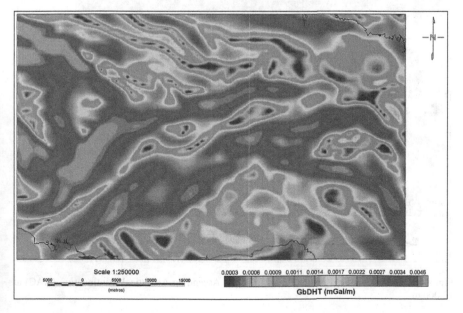

Fig. 4.2 Map of the total horizontal derivative of the gravitational field

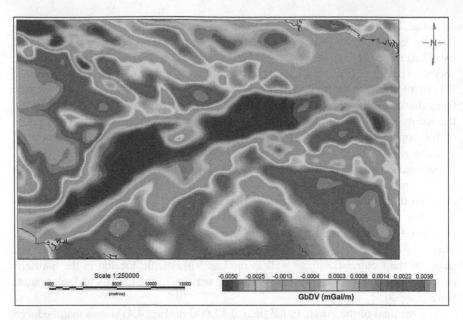

Fig. 4.3 Map of the first-order vertical derivative of the gravitational field

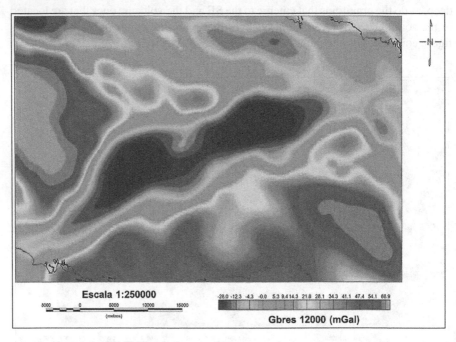

Fig. 4.4 Gravitational field residual map from the Ascending Analytical Continuation (AAC) at 12,000 m

The data resulting from the magnetic surveys in this region have allowed several authors, such as Rifá Hernández (2012) and Pardo Echarte (2020), to study the distribution of OA and CVA in the Central Basin region. It is known that most of the known reservoirs are producers in tuffs and that many accumulations and manifestations of hydrocarbons are associated with magnetic maxima. Examples of the above are the Jatibonico, Catalina and Cristales oilfields.

The magnetic interpretation carried out in the research was made from the Grid of the anomalous magnetic field at a scale of 1:50,000 (Mondelo Diez et al. 2011) with a mean square error of ± 7.67 nT. In the same way as in the gravitational field, the anomalies associated with ophiolites and volcanic rocks correspond to maxima; the ultrabasites being the ones that provide the most noticeable response. The maps of the pole-reduced magnetic field (RP) (Fig. 4.5) as well as its first-order vertical derivative (Fig. 4.6) are given below.

In the map of the reduced to the pole magnetic field, several positive anomalies are observed. Those located to the south and southeast are associated with volcanic rocks belonging to the CVA and large bodies of the Gabbro-Plagiogranitic Complex (GPC), while the positive anomalies located to the northwest are associated with folded bodies of the Zaza Terrain. To the north, a minimum elongated from west to east is observed in Fig. 4.5; it is considered to be produced by the Remedios TSU,

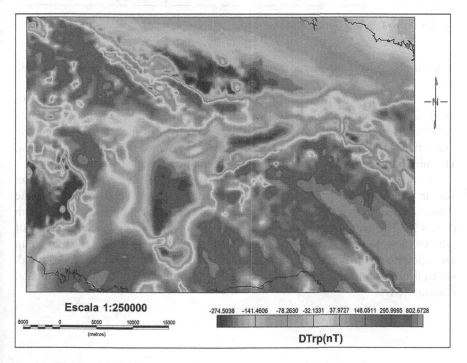

Escala 1:250000

5000   0   5000   10000   15000
(metros)

-274.5038 -141.4606 -78.2630 -32.1331 37.9727 148.0511 295.9995 802.6728

DTrp(nT)

Fig. 4.5 RP Magnetic field map

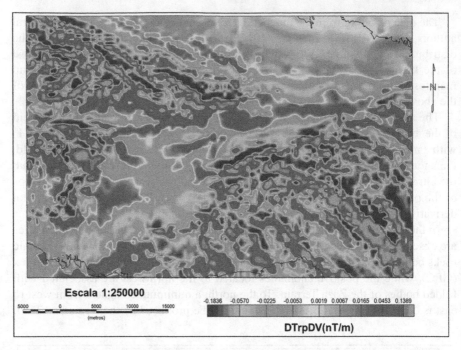

Fig. 4.6 Map of the first vertical derivative of the RP magnetic field

which is also seen in the Morón Norte 1 well, which cuts this stratigraphic tectonic unit.

The first vertical derivative of the pole-reduced magnetic field (Fig. 4.6) shows the distribution of ophiolitic bodies within the basin, as well as chains of maxima associated with CVA and OA rocks in the western part of the study area, and the rocks of the GPC in the eastern part, where both are oriented from west to southeast. The minimums are associated with the synorogenic sediments present in the area.

The residual magnetic field at 12,000 m shows that the large bodies of the Zaza Terrain located to the west extend to more than 12 km in depth, as well as the granitoids that outcrop to the SE and the volcanic ones that are located to the south (Fig. 4.7). The central zone presents a great minimum, indicating the presence of syn- and postorogenic sediments. This minimum is divided in two by the maximum corresponding to the uplift of the volcanics in the area of the Jatibonico oilfield, which is also assumed to be linked to a rise of the carbonates of the Camajuaní and Placetas TSUs.

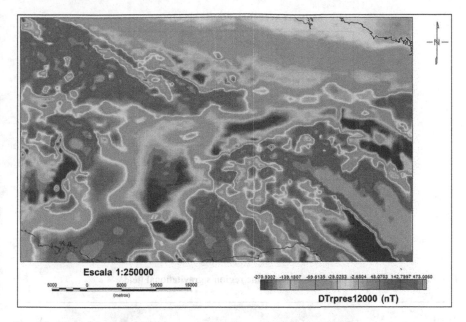

Fig. 4.7 Map of the residual RP magnetic field from the AAC at 12,000 m

4.3.2 Modeling the Potential Fields

Prior to the development of 2D physical-geological modeling, the spectral composition of the main density limits in the region was explored using the radial average power spectrum of the gravitational field (Fig. 4.8). Three main limits were established: the top of the basement (~12 km); the carbonate top of the NACM (~6 km) and; the top of the volcanics (effusive) (~4 km). A last level, ~1 km, characterizes noise.

The 2D physical-geological models, whose location is shown in Fig. 4.9, were made for a maximum depth of 12 km using the residuals of the potential fields at that depth since, according to Arriaza (1998), the top of the basement is located at a depth that varies between 8 and 10 km. The algorithm applied in the modeling was that of Talwani et al. (1959), Talwani and Heirtzler (1964) and other complementary ones used by the geophysical software (GMSYS-Oasis Montaj) (Talwani and Heirtzler 1964; Talwani et al. 1959). As mooring data, the data of the wells that were close to or on the study profiles and the interpretation of several seismic lines were taken.

The density values for the modeling were obtained from a statistical analysis carried out with the digital data of the wells drilled in the Central Basin, belonging to a database of the Centro de Investigación del Petróleo (CEINPET) (Colectivo de Autores 2005). In the analysis, it was found that the different geological units behaved as independent populations (Table 4.1).

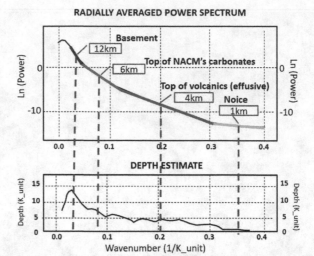

Fig. 4.8 Radial average power spectrum of the region's gravitational field

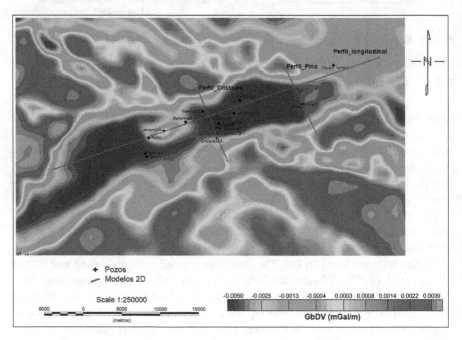

Fig. 4.9 Location of the 2D physical-geological models of the Central Basin on the map of the first vertical derivative of the gravitational field

Table 4.1 Statistical parameters for each of the geological objects (density in t/m^3)

Statistical parameters		Geological objects						
		Postorogenic sediments	Synorogenic sediments	Tuff	Effusive	Ophiolites	Carbonates	Basement
Number of samples			344	559	308	53	344	52
Arithmetic average		2.460	2.271	2.225	2.404	2.516	2.658	2.781
Variance		0	0.069	0.051	0.050	0.055	0.023	0.003
Confidence interval	Lower limit	2.434	2.235	2.201	2.372	2.433	2.637	2.760
	Upper limit	2.486	2.308	2.250	2.437	2.599	2.679	2.802

During the modeling, it was found that the magnetic data did not fit satisfactorily, so it was decided to show only the result of the 2D gravitational models. However, the RP magnetic field map data (residual at 12,000 m) were of vital importance in locating the structural uplifts of the Zaza Terrain. They also provided relevant results in the mapping of the Tectonostratigraphic Units present.

The mean square error obtained in the gravitational field modeling was 0.6 mGal, and this was not tried to decrease further in order not to affect the geological coherence of the models.

According to Cruz Orosa (2012), variations in the structural pattern and sedimentation in the synorogenic basins provide essential information to define the structural evolution of each territory. When analyzing the southwest and northeast sectors of the Central Basin, it is seen that there is a spatial variation in the structural pattern (Fig. 4.10). The southwest of the basin resembles a semigraben developed in a transtensional regime, while the northeast sector appears as a triangle zone (push-down) developed in a transpressive regime. The thickness of the NACM, according to the results of the present modeling, decreases from north (5 km) to south (1.3 km) and, conversely, the one of Zaza Terrain.

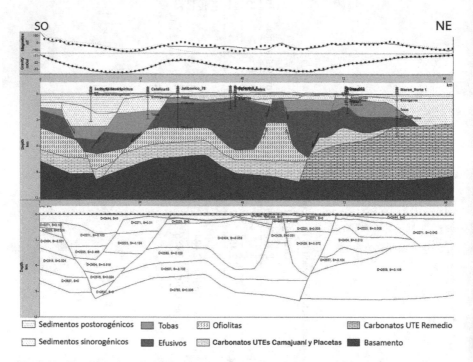

Fig. 4.10 2D physical-geological model longitudinal to the Central Basin, from the residual at 12,000 m of the gravitational field. Postorogenic sediments; synorogenic sediments, in light green, tuffs; in dark green, effusive; ophiolites; carbonates from the Camajuaní and Placetas TSUs; carbonates from Remedios TSU; in brown, basement (Color figure online)

According to Cruz Orosa (2012), the characteristics, age and lateral variations of the sedimentary fill have made it possible to detail the evolution of the convergence and propose that the accretion of the different segments of the orogen to the North American Plate migrated to the east. The synorogenic sedimentary fill of the basin indicates rapid development and a high rate of subsidence. These sediments include olistostromic and turbiditic series that present large lateral changes of facies and thickness.

Figure 4.11 shows the central part of the basin. The model with a SE- NO direction crosses the Cristales oilfield, cutting the Cristales 54, Cristales 31, Cristales Norte 1, Reforma 6 and Marroquí 2 wells. The area is divided by the following faults: Cristales, located to the SE of the oilfield, and Zaza-Tuinicú. The first behaves like a passing fault that generates other secondary faults, all of which facilitated the migration of hydrocarbons. The Zaza-Tuinicú fault was active until the Eocene, this is corroborated by the sedimentary layers present in it. This fault appears to be active again during the Oligocene and Miocene, but the absence of a sedimentary record makes its ratification impossible. The Cristales fault, on the other hand, begins its activity later, until the Early Miocene. The activity of the Zaza-Tuinicú and Cristales

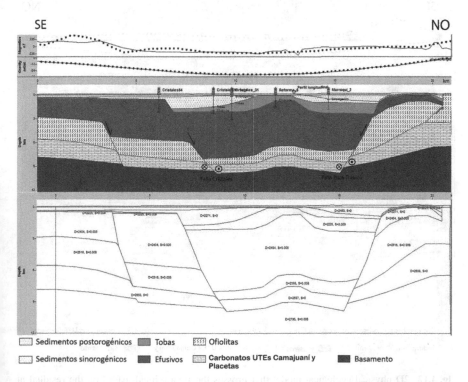

Fig. 4.11 2D physical-geological model that crosses the Cristales oilfield, from the residual at 12,000 m of the gravitational field. Postorogenic sediments; synorogenic sediments, in light green, tuffs; in dark green, effusive; ophiolites; carbonates from the Camajuaní and Placetas TSUs; in brown, basement (Color figure online)

faults gave rise to the occurrence of other faults that have produced a system of blocks in the basin. According to the results of the present modeling, the top of the NACM in this sector of the basin ranges between 5.55 and 6.6 km in Jatibonico-Cristales and 6.2 km in Catalina.

Figure 4.12 shows the 2D physical-geological model based on the gravitational field that crosses the Pina oilfield. The following were used as primary data: the residual at 12,000 m from this field, the petrophysical data from the wells Pina Norte 1, Pina 31, Pina 30, Pina 24, Pina 2 and Pina 3; and the depth of the top of the NACM rocks (carbonates from the Camajuaní and Placetas TSUs), taken from the research carried out by Rifá Hernández (2012). Based on the interpretation of several seismic lines and potential fields, the aforementioned author concluded that the depth of the NACM in this sector ranges from 3 to 3.8 km with an ascent toward the northwest. The values obtained from the present 2D modeling range between 2.98 and 4.3 km.

The postorogenic and synorogenic sediments have a total thickness in the Pina deposit of 1 km, although this thickness decreases to the NW. Unlike the previous models (Figs. 4.9 and 4.10), the carbonate thicknesses of the Camajuaní and Placetas

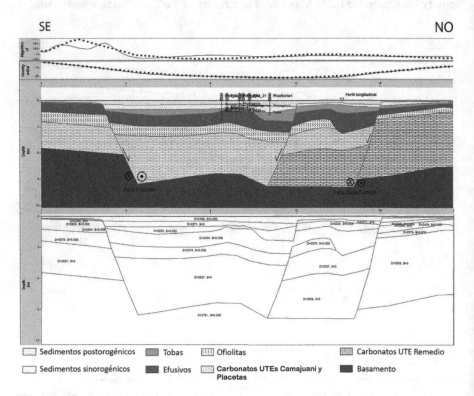

Fig. 4.12 2D physical-geological model that crosses the Pina oilfield, based on the residual at 12,000 m from the gravitational field. Postorogenic sediments; synorogenic sediments, in light green, tuffs; in dark green, effusive; ophiolites; carbonates from the Camajuaní and Placetas TSUs; carbonates from Remedios TSU; in brown, basement (Color figure online)

TSUs present thicknesses that exceed 3 km. This reinforces the hypothesis of the best prospects for finding conventional oil from these TSUs in the Pina oilfield sector.

4.4 Conclusions

- Geological and petrophysical data, seismic data and potential fields of the north-eastern region of the Central Cuba Basin were evaluated, in order to integrate them in the preparation and interpretation of three 2D physical-geological models of potential fields: one that is longitudinal to the basin and two that cut to the Cristales and Pina oilfields, respectively.
- As a result from the 2D physical-geological models, the hypothesis of the existence in the whole basin of carbonates from the NACM, considered as source rock, is validated. According to the models, the top of these rocks is located, at the Pina sector, between 2.98–4.3 km, while at the Jatibonico-Cristales and Catalina sectors, they range between 5.55–6.6 km and 6.2 km, respectively. In addition, their thickness decreases from north (5 km) to south (1.3 km) and, conversely, the one of Zaza Terrain. This reinforces the hypothesis of the best prospects for finding conventional oil and gas from the Camajuaní and Placetas TSUs in the Pina sector.

Acknowledgements The authors wish to thank the Centro de Investigación del Petróleo for allowing the use of the information necessary for the development of the research. Also, they want to thank Dr. Juan Guillermo López Rivera, Dr. Rolando de Armas Novoa, Dr. Ramón González Caraballo, Dr. José A. Díaz Duque, Dr. Evelio Linares Cala, Dr. Emilio Escartín Sauleda and Dr. Héctor Fernández Núñez, for his always timely collaboration.

References

Arriaza GL (1998) Nuevos enfoques en la interpretación y procesamiento de las ondas refrac-
tadas para el estudio del Basamento en Cuba: La Habana, Cuba. Doctoral thesis in Sciences,
unpublished., Instituto Superior Politécnico José Antonio Echeverría
Batista Rodríguez JA, Pérez Flores MA, Blanco Moreno J, Camacho Ortegón LF (2014) Structural
deformation in central Cuba and implications for the petroleum system: new insights from 3D
inversion of gravity data. Rev Mexicana Ciencias Geol 31(3):325–339
Cobiella Reguera JL (2017) Mapa tectónico del basamento de Cuba, escala 1:500,000. Unpublished
report. Instituto de Geología y Paleontología, La Habana, Cuba
Colectivo de Autores (2005) Base de datos de propiedades físicas de las rocas en la Cuenca Central
de Cuba por datos de pozos. (Unpublished report). Centro de Investigaciones del Petróleo, La
Habana, Cuba
Colectivo de autores (2008) Mapa Digital de las Manifestaciones de Hidrocarburos de la República
de Cuba a escala 1:250,000 (Unpublished report). Centro de Investigaciones del Petróleo, La
Habana, Cuba

Colectivo de autores (2009) Mapa Digital de los Pozos Petroleros de la República de Cuba a escala 1:250,000 (Unpublished report). Centro de Investigaciones del Petróleo, La Habana, Cuba

Colectivo de autores (2010) Mapa Geológico Digital de Cuba a escala 1:100,000 (Unpublished report). Instituto de Geología y Paleontología, La Habana, Cuba

Cruz Orosa I (2012) Las cuencas sinorogénicas como registro de la evolución del Orógeno cubano: Implicaciones para la exploración de hidrocarburos. Docotoral thesis in Sciences, unpublished, University of Barcelona

Cruz Orosa I, Sàbat F, Ramos E, Vázquez Taset YM (2012a) Synorogenic basins of central Cuba and collision between the Caribbean and North American plates. Int Geol Rev 54:876–906

Cruz Orosa I, Vázquez Taset YM, Sàbat F, Ramos E, Bernaola G (2012b) Segmentation and welding of the Cuban Orogen. A discussion about the evolution of the NW-Caribbean. Terra Nova

García R, Valdés P (2004) Reporte de las investigaciones geológicas sobre las líneas sísmicas en el sector noroccidental de la Cuenca Central, Bloque 21. Provincia Ciego de Ávila (Unpublished report). Centro de Investigaciones del Petróleo, La Habana, Cuba

Hatten CW, Schooler OE, Giedt N, Meyerhoff AA (1958) Geology of central Cuba, eastern Las Villas and western Camagüey provinces. Archivo del Servicio Geológico Nacional, La Habana, Cuba

Ipatenko SP (1968) Informe sobre las investigaciones magnetométricas y gravimétricas en la provincia de Camagüey. Unpublished report, Archivo del Servicio Geológico Nacional, La Habana, Cuba

Kireev I (1963) Informe referente a la exploración gravimétrica de la Cuenca Central. Instituto Cubano de Recursos Minerales, La Habana, Cuba

Martínez Martínez Y (2005) Modelación 3D de datos Gravimétricos de la parte sur de la Cuenca Central. Graduation Thesis, unpublished. Instituto Superior Minero Metalúrgico de Moa "Dr. Antonio Núñez Jiménez", Cuba

Martínez Rojas E, Toucet S, Sterling N, Yparraguirre JL (2006) Informe sobre la reinterpretación geólogo-geofísica y evaluación estructural del Bloque 21. Reinterpretación sísmica terrestre 2D. Unpublished report. Centro de Investigaciones del Petróleo, La Habana, Cuba

Meyerhoff AA, Hatten CW (1974) Bahamas salient of North America: tectonic framework, stratigraphy, and petroleum potential. Am Asso Petrol Geol Bull 58:1201–1239

Meyerhoff AA, Hatten CW (1968) Diapiric structure in central Cuba. Am Assoc Pet Geol Mem 8:315–357

Mondelo F, Sánchez R et al. (2011) Mapas geofísicos regionales de gravimetría, magnetometría, intensidad y espectrometría gamma de la República de Cuba, escalas 1:2000,000 hasta 1:50,000. Unpublished reports. Instituto de Geología y Paleontología, La Habana, Cuba. Nro. TTP 617965

Pardo Echarte ME (2020) Cartografía geólogo-estructural y sectores perspectivos para hidro-carburos en Cuba Central a partir de métodos no-sísmicos de exploración. Geociencias UO 3(1):35–44.

Peña Reyna A (2005) Modelación 3D de datos gravimétricos del norte de la Cuenca Central. Graduation thesis, unpublished. Instituto Superior Minero Metalúrgico de Moa "Dr. Antonio Núñez Jiménez", Cuba

Peña Reyna A, Batista Rodríguez JA, Blanco J (2007) Nuevas regularidades estructurales de la Cuenca Central (Cuba) a partir de la interpretación cualitativa de datos gravimétricos. Minería Geología 23(1):20

Pindell JL, Kennan L, Maresch WV, Stanek KP, Draper G, Higgs R (2005) Plate-kinematics and crustal dynamics of circum-Caribbean arc-continent interactions. Tectonic controls on basin development in Proto-Caribbean margins, in Caribbean-South American plate interactions, Venezuela. GSA Special Papers 394:7–52

Rifá Hernández M (2012) Ubicación de los sectores elevados de los sedimentos del Margen Conti-nental en la Cuenca Central, Cuba. Master's of Sciences thesis, unpublished. Instituto Superior Politécnico José Antonio Echeverría

Rodríguez M, Domínguez R (1993) Informe sobre los resultados del levantamiento gravimétrico en Jatibonico-Pina-Esmeralda. Unpublished report. Archivo de la Empresa Nacional de Geofísica, La Habana, Cuba

Rodríguez M, Pról JL (1980) Informe sobre el levantamiento gravimétrico detallado del área Mayajigua-Morón. Unpublished report. Archivo de la Empresa Nacional de Geofísica, La Habana, Cuba

Rosencrantz E (1990) Structure and tectonics of the Yucatan Basin, Caribbean Sea, as determined from seismic reflection studies. Tectonics 9:1037–1059

Talwani M, Heirtzler JR (1964) Computation of magnetic anomalies caused by two-dimensional bodies of arbitrary shape. In: Computers in the mineral industries, Proceedings of the Third annual conference sponsored by Stanford University School of Earth Sciences and University of Arizona College of Mines. United States

Talwani M, Lamar Worzel J, Landism M (1959) Rapid gravity computations for two-dimensional bodies with application to the Mendocino Submarine fracture zone. J Geophys Res 64(1):49–59

Chapter 5
Evaluation of Areas that Meet the Necessary Geological-Petroleum Premises for the Presence of Large Accumulations of High-Quality Oil

Orelvis Delgado López, Juan Guillermo López Rivera,
José Orlando López Quintero, and Zulema Domínguez Sardiñas

Abstract Cuba has been producing oil since 1936. The first fields produced good quality crude but with little production due to bad reservoirs (ophiolites). Later, better reservoirs were discovered (Mesozoic carbonates) with large resources but with poor quality oil. So far, no large reserves of high-quality oils have been discovered in Cuba. The objective of this study has been to evaluate areas that meet the petroleum-geologist premises for the presence of large accumulations of high-quality oil. For that reason, 2038 rock samples (for rock-eval studies), 207 oil samples (for physicochemical and biomarkers analysis), and 27 organic extracts from source rock (for oils, source rock correlations) were used. The petroleum systems exploratory method was followed. Only those areas with high thermal evolution oils were taken into account (Exploratory blocks 21–23, Block 7, Block 6, Block 13 and Blocks 17–18). It is concluded that the presence of families I and II of Cuban oils indicate the presence of the J_3-K_1 tectonic sheet; Family III of Cuban oils, indicates the presence of the K_1-K_2 tectonic sheet. The premises that an area must meet for the existence of large accumulations of high-quality oil are: oil with high thermal evolution, rich in sulfur and protected from biodegradation, Veloz Group type reservoirs. Of the areas evaluated, Block 21–23 is the one that meets all the premises for the presence of large accumulations of high-quality oil. In the rest, there is variation in thermal maturity, increasing the risk of finding deep accumulations with high-quality

O. Delgado López (✉) · J. G. López Rivera · Z. Domínguez Sardiñas
Centro de Investigación del Petróleo, Churruca, No. 481, Vía Blanca y Washington, 12000 El Cerro, La Habana, CP, Cuba
e-mail: orelvis@ceinpet.cupet.cu

J. G. López Rivera
e-mail: juang@ceinpet.cupet.cu

Z. Domínguez Sardiñas
e-mail: zulds@ceinpet.cupet.cu

J. O. López Quintero
Centro Politécnico del Petróleo, Vía Blanca y Ave de los Mártires, 11600 Regla, La Habana, CP, Cuba

© The Author(s), under exclusive license to Springer Nature Switzerland AG 2022
M. E. Pardo Echarte et al., *Geological-Structural Mapping and Favorable Sectors for Oil and Gas in Cuba*, SpringerBriefs in Earth System Sciences,
https://doi.org/10.1007/978-3-030-92975-6_5

commercial crude, and also, presence of reservoirs that have not demonstrated large reserves (Camajuaní Tectonostratigraphic Unit).

Keywords Non-seismic methods of oil and gas exploration · Favorable and perspective sectors for hydrocarbons · Petroleum systems · Cuban oil families · Cuba

Abbreviations

NACM	North American Continental Margin
CVA	Cretaceous Volcanic Arc
CUPET	Unión Cuba Petróleo
NCOB	Northern Cuban oil belt
Fm.	Geological formation
TSU	Tectonostratigraphic unit
PVT	Pressure volume temperature
°API	American Petroleum Institute, oil density measurements unit

5.1 Introduction

Oil has been produced in Cuba since the beginning of the last century (1936–1942) when some oil fields were developed that produced good quality crude, such as the cases of Motembo (64.5 °API, 0.004% S), Jarahueca (42.2 °API, 0.6% S) and Bacuranao-Cruz Verde (28.3 °API, 1.0% S). All these oil fields had in common that their production came from ophiolites, rocks that do not have good reservoir properties, so their production was quite limited and with few hydrocarbon reserves. Subsequently, the development of petroleum exploration in Cuba allowed the discovery of oil fields in sedimentary rocks, associated with the North American Continental Margin (NACM) with large hydrocarbon reserves such as the cases of Varadero, Boca de Jaruco, Puerto Escondido, Seboruco and Santa Cruz. With these discoveries, a very prospective area for petroleum exploration emerged in the country known as the North Cuban Oil Belt (NCOB). However, this area has the problem that, although there is some production of light and medium oil in it, most of the crude produced in it is classified as heavy and extra heavy with high sulfur content. Also, at the beginning of the 90s of the last century, some oil fields were discovered in reservoirs associated with the Cretaceous Volcanic Arc (CVA) with productions of good quality oil, such as Pina (31.4 °API, 1.71% S) and Brujo (45.9 °API, 0.53% S). But like the first fields discovered in Cuba, daily productions and reserves are low, showing that only in reservoirs associated with sedimentary rocks of the NACM the existence of large hydrocarbon reserves was guaranteed.

Table 5.1 Classification of oil according to its density and sulfur content

°API	Classification	Sulfur content (%)	Classification
> 31.1°	Light	< 0.5	Low sulfurous
22.3–31.1°	Medium	0.5–1.5	Medium sulfurous
10–22.3°	Heavy	1.5–3.0	Sulfurous
< 10°	Extra heavy	> 3.0	High sulfurous

Several have been the works carried out and attempts to locate large oil fields of light oil in Cuba, but to date it has not been achieved. Sometimes for geological reasons and other times for technological problems. The objective of this research was to evaluate areas of Cuba that meet the geological-petroleum premises necessary for the presence of large accumulations of high-quality oil.

5.2 Theoretical Framework

In this section, a series of concepts on organic geochemistry are exposed that will help to understand the results achieved.

5.2.1 Commercial Quality of Oil

From an exploratory point of view, there are two parameters that characterize the commercial quality of oil: the density (expressed in °API) and the sulfur content. Table 5.1 shows the crude oil classification used on the basis of their commercial quality.

5.2.2 Geological Processes that Determine the Quality of Oil

The main geological processes that determine the commercial quality of crude oil can be divided into two big groups on the basis of when they take place:

- **Primary processes**: They determine the commercial qualities of oil from the moment of its generation and are determined by the organic and thermal characteristics of the source rocks, that is, they occur at source rock level. These are: 1. The Origin and 2. Thermal maturation.

    1. Origin: It refers to the type of organic matter contained in the source rock that depends on the depositional environment. In a general sense, all oils contain sulfur compounds, but they predominate in carbonate source rocks more than in clastic

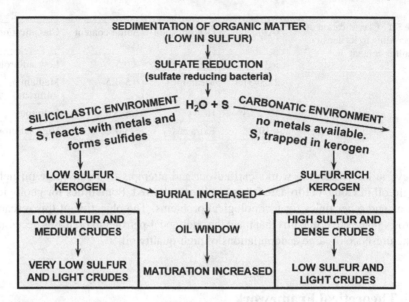

Fig. 5.1 Primary processes that determine the commercial qualities of oil

ones. During the sedimentation processes of organic matter, sulfates (SO_4^{2-}) are transformed into sulfides (S^{2-}) through the action of sulfate-reducing bacteria. In clastic environments, sulfur combines with existing metallic elements to form sulfides, thus remaining in the rock. But in the case of carbonate environments, where metals are not abundant or absent, the sulfur released from sulfates is incorporated directly into kerogen (Fig. 5.1).

2. Thermal Maturation: Maturation is a thermal process that guarantees that a source rock can transform kerogen into hydrocarbons and that also determines its initial physicochemical properties. Once the source rock reaches the appropriate temperature ($\approx 100°$ C), it begins to generate oil, and as the temperature increases, increasingly lighter hydrocarbons are generated from the kerogen and heavier oil already generated and not expelled (Fig. 5.1).

- **Secondary Processes**: They refer to alterations suffered by oil during migration or after accumulation. These tend to decrease the oil quality and its economic value, and occur at the reservoir level: Biodegradation, water washing, evaporative and gravitational fractionation and deasphaltinization. The most common of these process in Cuba are, 1. Biodegradation (Fig. 5.2) and 2. Fractionation (Fig. 5.3).

1. Biodegradation: It consists in the alteration of oil by the bacteria action during migration, accumulation or in oil seeps. Biodegradation of oil only occurs at temperatures < 80°C, shallow depths and conditions where groundwater with dissolved oxygen is available to aerobic bacteria. Microorganisms typically degrade oil by first attacking the less complex, hydrogen-rich compounds, such as n-paraffins and isoprenoids, causing a decrease in the API gravity of the oil and an increase in sulfur content, due to the selective removal of the oil saturated and

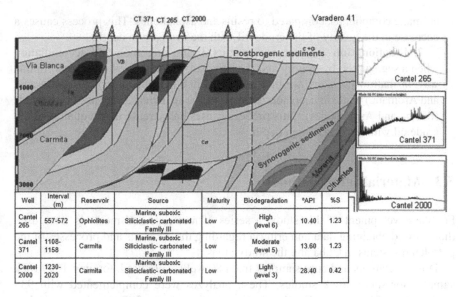

Well	Interval (m)	Reservoir	Source	Maturity	Biodegradation	°API	%S
Cantel 265	557-572	Ophiolites	Marine, suboxic Siliciclastic- carbonated Family III	Low	High (level 6)	10.40	1.23
Cantel 371	1108-1158	Carmita	Marine, suboxic Siliciclastic- carbonated Family III	Low	Moderate (level 5)	13.60	1.23
Cantel 2000	1230-2020	Carmita	Marine, suboxic Siliciclastic- carbonated Family III	Low	Light (level 3)	28.40	0.42

Fig. 5.2 Geological scheme of the Cantel oil field, showing the influence of the biodegradation process on the variation in oil quality

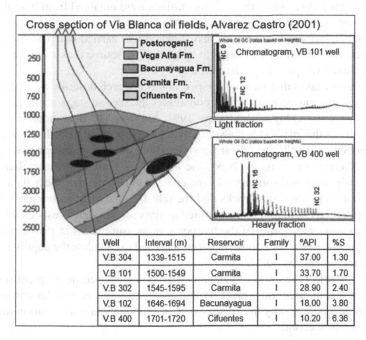

Well	Interval (m)	Reservoir	Family	°API	%S
V.B 304	1339-1515	Carmita	I	37.00	1.30
V.B 101	1500-1549	Carmita	I	33.70	1.70
V.B 302	1545-1595	Carmita	I	28.90	2.40
V.B 102	1646-1694	Bacunayagua	I	18.00	3.80
V.B 400	1701-1720	Cifuentes	I	10.20	6.36

Fig. 5.3 Geological scheme of the Vía Blanca oil field, showing the influence of the fractionation process on the variation in oil quality

aromatic compounds, compared to resins and asphaltenes. This process causes a decrease in the quality of the crude oil with the decrease in depth.

2. Fractionation: This process occurs during migration and accumulation when a segregation by gravity and therefore a fractionation of oil occurs. In this way, in shallow horizons, we find crude oils rich in light compounds (Saturated and Aromatic) while in deep horizons, oil shows an increase in heavy compounds (Resins and Asphaltenes). This process causes an increase in the quality of the crude oil with the decrease in depth.

5.3   Materials and Methods

For the development of this study, a series of materials and methods were used that allowed obtaining data and criteria regarding the elements and processes of the petroleum systems in Cuba and the study areas.

Drilling cuttings and core samples from 80 wells were used, totaling 1099 rock samples for source rock studies. These analyzes were complemented with 939 samples from 71 outcrops. For the characterization of oils, 207 samples were taken from oil fields and isolated producing wells. The drill cores used were located in the CUPET core store, while the drilling cuttings were obtained from both the core store (old wells) and wells drilled in the last five years. The oil samples used were all sampled in the last years before this investigation was carried out. To establish the source rock-oil correlations, 27 organic extracts from source rock of different ages and in immature samples were carried out.

The scientific tasks that were developed in this research describe the exploratory method of petroleum systems, which constitutes the second stage of the petroleum exploration process (Magoon and Dow 1994; Magoon and Beaumont 1999). The starting point of this method is the study by biomarkers of all the oil shows present in an area. Subsequently, the oils are correlated to establish groups or families that indicate different source rocks. Next, the defined oil families are correlated with the previously identified source rocks. In order to obtain the necessary data for the characterization of the source rocks and the oils, the analytical techniques of rock-eval pyrolysis and chromatography coupled to mass spectrometry, respectively, were used. In addition, the results of the temperature measurements in petroleum wells were used to determine the geothermal gradients and to calculate the depths at which the temperature reached 80 °C.

In order to select the areas to be evaluated, geological-petroleum premises were established that an area must meet so that there are large accumulations of good quality oil. In this way, only those areas with oils of high thermal evolution (mature) were taken into account.

5.4 Results

This section presents the results achieved in the study of oil, source rocks and the correlations between them; in addition, on the reservoirs and geological events that influenced the processes of the petroleum systems identified in Cuba.

5.4.1 Oils

Biomarker studies carried out to date in Cuba demonstrate the existence of three genetic families of petroleum (Pascual et al. 2001). An example of the relationships used to define these families is shown in Fig. 5.4, in which it is observed that according to the characteristics of the source rocks that generated them, Family II has the most carbonate origin, and Family III is the opposite. The source rock of Family I is carbonate but with a contribution of siliciclastics.

On this basis and according to what is expressed in Fig. 5.1, families I and II are generated by source rocks with high sulfur contents in their kerogens, while the source rock of Family III does not have a sulfur source.

Figure 5.5 shows a summary of the main characteristics of the three families from the point of view of their origin. These genetic characteristics mean that for the same levels of thermal maturity of the source rock, Family III oils have better commercial qualities than those of Families I and II. Figure 5.5 shows that the oil fields associated with Family I predominate in the western part of Cuba, while Family II does so in the central portion. Family III are present throughout the island but in lesser volume.

Fig. 5.4 Biomarker relationships (Hopanes) showing origin of the Cuban crude families. In both axes, the carbonate origin increases toward the extremes

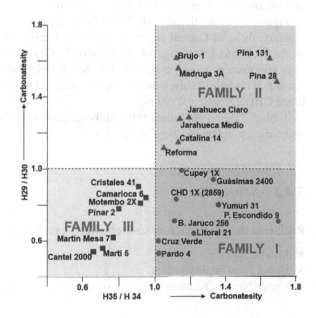

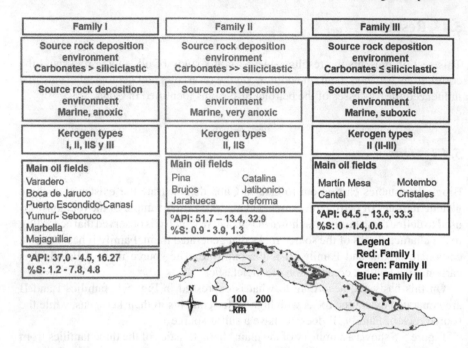

Fig. 5.5 Oil families identified in Cuba, main genetic characteristics (origin) and geographic distribution

This geographical distribution causes that the commercial accumulations of each family to have different thermal evolutions (Fig. 5.6), indicating different levels of maturation of the source rocks that generated them.

Figure 5.6 shows that the NCOB oil fields are the least thermally evolved, on the contrary, those of Central Basin are the most mature. Some wells, such as Mariel Norte 1X, Pardo 4, Madruga 3A, Motembo 2X, Cayajabos 3, Pacheco 2 and Chacón 2 are also mature, as well as the oil seeps San Gabriel and Mina Talaren. It should be noted that the most evolved crude from the thermal point of view is the one studied in the CHD 1X well at 2899 m.

Table 5.2 shows the names of the oil fields and isolated wells with their respective codes used to represent them in Fig. 5.6.

5.4.2 Source Rocks

Rock-eval studies reveal four stratigraphic intervals with source rock properties:

- Middle Jurassic
- Upper Jurassic, Oxfordian
- Upper Jurassic Kimmeridgian—Lower Cretaceous Barremian

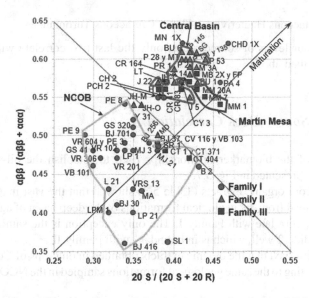

Fig. 5.6 Biomarker relationships (Steranes) showing the thermal evolution of Cuban crudes

Table 5.2 Codes of the oil fields, wells, oil seeps used to show levels of thermal evolution of Cuban crude oils

Oil field or well	Cod	Family	Oil field or well	Cod	Family	Oil field or well	Cod	Family
Basilio 2	B	I	Pacheco 2	PCH	I	Mina Desengaño	MD	II
Boca de Jaruco	BJ		Pardo 4	PA		Mina Talaren	MT	
Cayajabo 3	CY		P. Escondido	PE		Paraíso 1	PR	
Chacón 2	CH		San Lázaro	SL		Pina	P	
Chacón 1X	CHD		Santa Rita	SR		San Gabriel	SG	
Cruz Verde	CV		Vía Blanca	VB		Cristales	CR	III
Guásimas	GS		Yumurí	Y		Cantel	CT	
La Tomasita	LT		Brujo	BU	II	Floro Pérez	FP	
Litoral	L		Catalina	CA		Holguín	H	
Majaguillar	MJ		Jatibonico	J		Motembo	MB	
Mariel Norte 1X	MN		Jarahueca	JH		Martín Mesa	MM	
Mina Manuela	MA		Madruga 3A	M				

- Lower Cretaceous Hauterivian—Upper Cretaceous Turonian.

Of these four levels of source rocks, only the last two correlate with the three families described above.

5.4.3 Oil-Source Rock Correlation

An example of the biomarker relationships used to establish the oil-source rock correlations is presented in Fig. 5.7.

The data from organic extracts (Table 5.3) showed that the vast majority of the samples, obtained from the geological formations of the deep basin of age $J_3{}^{km}$-$K_1{}^{ba}$ in the NCOB, correlate with Family I. The only exception is the sample from the Puerto Escondido 2 well, which is in boundary with Family II.

It is also observed that the organic extracts obtained in outcrops of Central Cuba, and corresponding to the same geological formations sampled in the NCOB, correlate with Family II.

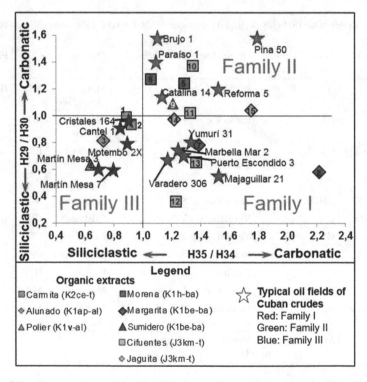

Fig. 5.7 Oil– source rock correlation for Cuban samples, according to biomarker relationships (Hopanes). The code for the extract samples appears in Table. The stars meaning typical Cuban oil fields (Red, Family I; Green, Family II; Blue, Family III) (Color figure online)

Table 5.3 Samples of organic extracts obtained from Cuban source rocks

Code	Samples	Well or Locality	Interval (m)	Formation	Location
1	940,987,018	Guásimas 41	1948–1952	Carmita (K_2^{ce-t})	NCOB
2	940,990,021	Boca de Jaruco 451	1251	Carmita (K_2^{ce-t})	NCOB
3	GQ160	Quarry Calienes	Outcrop	Alunado (K_1^{ap-al})	Central Cuba (Villa Clara)
4	940,989,932	Chacón 2	1384–1389	Polier (K_1^{h-al})	Western Cuba (Pinar)
5	AM-20	Loma Bonachea	Outcrop	Morena (K_1^{h-ba})	Central Cuba (Villa Clara)
6	AM-22	Loma Bonachea	Outcrop	Morena (K_1^{h-ba})	Central Cuba (Villa Clara)
7	AM-13	Aguada La Piedra	Outcrop	Margarita (K_1^{be-ba})	Central Cuba (Villa Clara)
8	AM-14	Aguada La Piedra	Outcrop	Margarita (K_1^{be-ba})	Central Cuba (Villa Clara)
9	EL-125–3–20	Las Lajas	Outcrop	Sumidero (K_1^{be-ba})	Western Cuba (Pinar)
10	AM-17	Loma Sin Nombre	Outcrop	Cifuentes (J_3^{km-t})	Central Cuba (Villa Clara)
11	940,990,456	Puerto Escondido 2	4002–4005	Cifuentes (J_3^{km-t})	FPNC
12	940,990,532	Varadero 201	2600–2604	Cifuentes (J_3^{km-t})	FPNC
13	940,990,155	Litoral Pedraplén Mar 1	2105–2110	Cifuentes (J_3^{km-t})	FPNC
14	GQ180	Loma Las Azores	Outcrop	Jaguita (J_3^{km-t})	Central Cuba (Villa Clara)
15	GQ190B	Aguada La Piedra	Outcrop	Jaguita (J_3^{km-t})	Central Cuba (Villa Clara)

This data indicates facial variations of the source rocks of the J_3^{km}-K_1^{ba} interval that causes it to generate Family I oils in the NCOB and Family II in Central Cuba. These facies variations of source rocks are very common in the mega-basin of the Gulf of Mexico (Sofer 1988; Requejo and Halpern 1990; Thompson 1990; Rocha Mello and Trindade 1996).

Finally, Fig. 5.7 shows that the Lower Cretaceous Valanginian-Upper Cretaceous Turonian age source rocks correlate with Family III, highlighting that the organic extract of the Polier Formation has more similarity with the crude ones of Martín Mesa, while the organic extract of the Carmita Formation has better correlation with the Cristales and Cantel oil fields.

Table 5.4 shows examples of the influence of primary processes (origin and

Table 5.4 Example of the influence of primary processes on the quality of Cuban oil

Oils	Origin	Source rock age	Family	Maturity	°API	%S
Varadero 102	Anoxic Marine Carbonate	Jurassic km-Cretaceous ba	I	Low	8.3	5.42
Pardo 4	Anoxic Marine Carbonate	Jurassic km-Cretaceous ba	I	High	35.0	1.14
Cantel 2000	Marine Sub oxic Siliciclastic-carbonated	Cretaceous ce-t	III	Low	28.4	0.42
Motembo 2X	Marine Sub oxic Siliciclastic-carbonated	Cretaceous ce-t	III	High	58.8	0.04

Note that, in the case of Family I, carbonate origin and therefore source rock rich in sulfur (Fig. 5.1), the oils are extra heavy and highly sulfurous in the case of the area with a low level of maturation (Fig. 5.6) and that they can become light and moderately sulfurous when the source rock reaches high levels of maturity. Family III from the genetic point of view is favored by the absence of sulfur in the source rock that gives it origin, highlighting that for high maturation they are very light crude. What most distinguishes Family III is the low sulfur content, even in the most biodegraded oils (Cantel 256), that it does not exceed 1.5%

thermal maturity) on the commercial quality of Cuban oil. It presents a comparison between crude oils from Families I and III in areas where the source rocks that generated them reached different levels of thermal maturity.

5.4.4 Reservoirs

The production data indicate that there are various levels of reservoirs in Cuba (Valladares et al. 1996). The vast majority of them are known in the NCOB, while those related to the CVA dominate the known oil fields in the Central Basin. In Fig. 5.8, a generalized geological column of the NCOB is shown to highlight the age of the reservoirs from which the production is obtained. It should be noted that the major volumes are obtained from Veloz Group (J_3^{km}-K_1^{ba}), which groups together the Cifuentes, Ronda and Morena formations (Fms.), thus constituting the main proven reservoir in Cuba.

There are several geological processes that conditioned Veloz Group to store the largest hydrocarbon resources. They can be divided into sedimentary and tectonic.

The carbonate marine sedimentation environments associated with the NACM, as well as anoxic and low-energy conditions ensured that:

- Veloz Group and the contemporary sediments of Cuba and the Gulf of Mexico constitute the main source rock in the region.
- The oil generated by Veloz Group is rich in sulfur because it is a source rock originated in a carbonate environment (Figs. 5.1 and 5.4). This hydrocarbon,

Fig. 5.8 Levels of reservoirs present in the NCOB

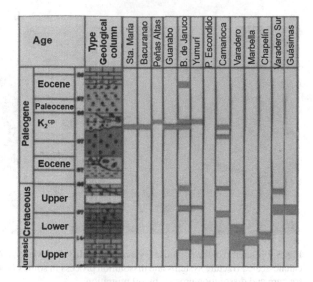

when migrating, reacts with water and forms sulfuric acid, a corrosive agent that dissolves carbonates.

- The carbonates of the Veloz Group, from the genetic point of view (clean carbonates), have a brittle behavior.

The tectonic processes that occurred during the Cuban orogeny, particularly the thrust, guaranteed the occurrence of the processes of the petroleum system related to Veloz Group:

- The appearance of two tectonic sheets that involve Mesozoic sediments of different ages, each of which is made up of several sheets.
- The Veloz Group is found in the lower tectonic sheet, Upper Jurassic Kimmeridgian to Lower Cretaceous Barremian age and several sheets, sometimes separated by synorogenic sediments of Paleocene-Eocene age (Vega Alta Fm. in the case of Veloz Group). This tectonic sheet identified in the Habana-Varadero region by wells is similar to the Rosario Sur described in outcrops by Pszczolkowski (1999).
- Creation of secondary porosity in the fragile carbonates of Veloz Group. This sheet thrusted (Paleocene-Eocene, synorogenic sediments age) when these carbonates were well consolidated, so these had a brittle behavior, giving rise to various fracture systems. The high fracture densities recorded in wells of the NCOB and outcrops (Fig. 9a) prove it.
- The big angles of inclination toward the south of the sheets cause the same sheet to act as source rock and reservoir at the same time. As source rock at the depths where the generation window reaches and as a reservoir on intensely fractured thrust fronts (Fig. 5.10).
- Fractures originated by thrusting are usually sealed by secondary calcite, but the migration of acidic fluids (high sulfur content in oil) dissolves them more easily than the matrix (Fig. 9a and b), thus increasing the permeability of these reservoirs.

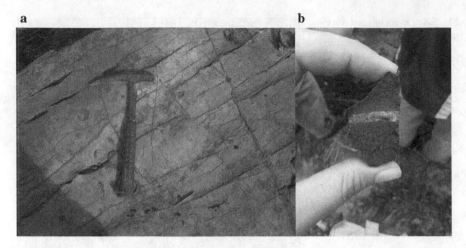

Fig. 5.9 Outcrop of the Artemisa Formation, locality: Los Maceos, Artemisa province. **a** Secondary calcite-sealed fracture systems and dissolution process in some of them. **b** Secondary calcite-sealed fracture and dissolution process by oil migration

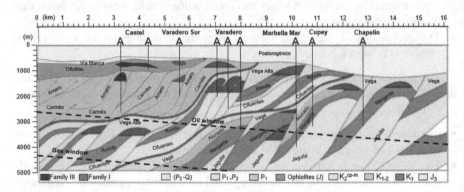

Fig. 5.10 Geologic scheme of Cantel-Varadero-Chapelín region, showing the active petroleum systems in that region and their relationship with the tectonic sheets identified by wells

On the contrary, the petroleum systems associated with the upper tectonic sheet (Cretaceous sediments of Hauterivian-Turonian age) were less favored by the same processes.

The marine sedimentation environments, with a large contribution of emerged lands by means of turbidites that oxygenated the basin, guaranteed that:

• The oil generated by the source rocks of this age is low in sulfur because it originates in environments with a high contribution of siliciclastic sediments (Figs. 5.1 and 5.4). This hydrocarbon, when migrating, does not form sulfuric acid and does not dissolve carbonate reservoirs.

Fig. 5.11 Outcrop of the Santa Teresa Formation in the La Sierra locality, Villa Clara province, showing plastic behavior of the rocks of the upper tectonic sheet in the NCOB [Rosario Norte, Pszczolkowski (1999)]

- Carbonates of this age, from the genetic point of view (carbonates with a high contribution of siliciclastics), have a more plastic behavior.

On the other hand, the tectonic processes that occurred during the Cuban orogeny, particularly thrust, caused that:

- The source rocks associated with the upper tectonic sheet, identified in the Habana-Varadero region by wells, and similar to Rosario Norte described on outcrops by Pszczolkowski (1999), are in a higher structural position than the lower sheet, which implies less volume of oil kitchen (Fig. 5.10) and therefore less volume of oil generated.
- Little creation of secondary porosity in the Lower Cretaceous Hauterivian-Upper Cretaceous Turonian age plastic carbonates. This tectonic sheet thrusted when these carbonates were poorly consolidated and therefore had a plastic behavior, so the greatest deformations were folding and not fractures. This aspect is visible in all the outcrops of the formations of this sheet (Fig. 5.11).

Based on the results presented above, a series of geological-petroleum criteria was elaborated for the petroleum systems present in the petroleum province of northern Cuba:

- The oils of families I and II are generated by source rocks of Upper Jurassic Kimmeridgian-Lower Cretaceous Barremian age, Veloz Group type and their synchronic sediments in the other Tectonostratigraphic Unit (TSU) identified in Cuba.
- This source rock level is contained in the lower tectonic sheet identified in the NCOB, which is similar to Rosario Sur described by Pszczolkowski (1999).
- The structural position of this sheet (deep) guarantees large volumes of source rock in the oil generation window.
- The brittle behavior of the rocks in this sheet ensured the formation of secondary porosity by fracturing during thrust.

- The high sulfur content in the oils generated by the source rock of this tectonic sheet led to the formation of acidic fluids that increased the permeability of carbonate reservoirs, type Veloz Group.
- Family III oils are generated by Lower Cretaceous Hauterivian-Upper Cretaceous Turonian source rocks, associated with the upper tectonic sheet, identified in the Habana-Varadero region by wells, and similar to the Rosario Norte described by Pszczolkowski (1999).
- The structural position of this sheet (shallower) causes little volume of source rock in the oil generation window.
- The plastic behavior of the rocks in this sheet made it impossible to form secondary porosity by fracturing during thrust.
- The low sulfur content in the oils generated by the source rocks of this tectonic sheet made the formation of acidic fluids impossible, therefore the permeability of the carbonate reservoirs was not increased.
- Family I oils have been found in reservoirs of the upper sheet ($K_1{}^h$-$K_2{}^t$) (Rosario Norte) but never crude oils of Family III have been found in the lower sheet ($J_3{}^{km}$-$K_1{}^{ba}$) (Rosario Sur). The cause must be the structural relationship between the two sheets.
- The presence of a specific family of oil in an area indicates the presence in the underground of a determined tectonic sheet according to the oil-source rock correlations:

    Families I and II indicate the presence of a lower tectonic sheet type Rosario Sur ($J_3{}^{km}$-$K_1{}^{ba}$).
    Family III indicates presence of upper tectonic sheet type Rosario Norte ($K_1{}^h$-$K_2{}^t$).

5.5 Discussion

Based on the results and criteria set out above, a number of geological-petroleum premises were established that an area must meet in order for there to be large accumulations of good quality oil:

- The oil must be of high thermal evolution resulting from a high level of thermal maturity of the source rock that generated it. This aspect must be verified by data from thermal biomarkers (Steranes or Diamandoids). It is not enough that the crude in an area has good quality, because there are other secondary processes that can cause the presence of light oils such as gravitational fractionation.
- Oil must be protected from biodegradation process. Reservoir temperatures above 80 °C. The reservoir Pressure Volume Temperature (PVT) is the ideal data, but temperature can be calculated from geothermal gradients in the area of interest.
- Oil must contain sulfur, to ensure the formation of acidic fluids that increase the permeability of carbonate reservoirs by dissolution. According to the studies carried out on Cuban crude oils, this condition is only met by families I and II,

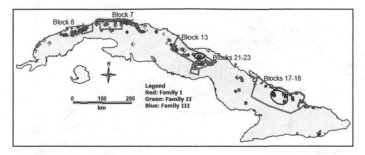

Fig. 5.12 Location of the areas studied to evaluate which ones comply with the geological-petroleum premises necessary for the presence of large accumulations of good quality oil

since Family III is very low in sulfur, even those generated by rocks with low levels of thermal maturity.

- The reservoir must be carbonates of the Upper Jurassic—Lower Cretaceous age, Veloz Group type, since it is the best play demonstrated in Cuba according to the proven reserves. This reservoir is located in the lower tectonic mantle of the NCOB (Rosario Sur, Pszczolkowski 1999) and also constitutes the main source rock of Cuba and the Gulf of Mexico. Its presence in the underground can be inferred by the presence of families I or II according to the source rock-oil correlations.

Next, the study carried out in several areas (Blocks 21–23, Block 7, Block 6, Block 13 and Blocks 17–18) following the order of analysis indicated by the geological-petroleum premises listed above is showed (Fig. 5.12).

5.5.1 Blocks 21–23

This area refers specifically to the Central Basin. Figure 5.6 shows that, from a regional point of view, the crudes from this area are the most mature in Cuba, since the other mature oils correspond to isolated wells (CHD 1X, Mariel Norte 1X, Pardo 4, Madruga 3A, Cayajabos 3, Pacheco 2 and Chacón 2), oil seeps (La Tomasita, San Gabriel, Mina Talaren, Floro Pérez and Holguín) or oil fields from different areas (Motembo and Martín Mesa). Note that all the oil fields and wells (Paraíso 1) in Central Basin show high levels of thermal evolution, even those of Family III, which, as explained above, is shallower than oils of families I and II. These data ensure that the possible accumulations of oil in depth, below the CVA and ophiolites, are of mature crude oil and therefore of high commercial quality (light and moderately sulfurous).

According to the values of the geothermal gradient in this area (21.1–39.3, average: 29.9 °C/km) (Fig. 5.13), the oil accumulations are protected from biodegradation (temperatures > 80 °C) from a depth of 2500 m.

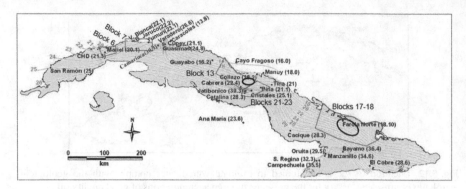

Fig. 5.13 Values and trends of geothermal gradients in Cuba. In magenta polygons, surveyed areas

Figures 5.4 and 5.5 show that the majority of the Central Basin oil fields belong to Family II of Cuban crude oils. This causes the sulfur content to be high, being in many cases above 1.00%, even in those not affected by biodegradation. When these oils undergo severe biodegradation, they can reach values of 3.89%. This is due to the high sulfur content in the source rock that generated them as they were formed in a highly carbonated environment (Figs. 5.1 and 5.2). However, the density expressed in °API is also high, and the majority classify as light oils in the case of those not affected by secondary processes. This is due to the fact that all the crude oils from this area have high levels of thermal evolution as explained previously (Fig. 5.6). Table 5.5 shows a statistic by oil field of the commercial qualities of these oils.

This sulfur content favors that, during migration and reaction with water, these crude oils form acidic fluids that dissolve the carbonate reservoirs that must be deep below the CVA and the ophiolites, thus increasing their permeability.

The fact that the Central Basin oil fields belong to Family II also indicates the presence of the Rosario Sur type lower tectonic sheet (J_3^{km}-K_1^{ba}), according to the oil-source rock correlations. This sheet, as already explained, constitutes the main exploratory play in Cuba (good porosity and permeability), which guarantees large volumes of accumulated oil. The fact that the oil accumulations, above this supposed sheet in depth, constitute oil fields, suggests large volumes accumulated in depth.

Table 5.5 Statistical data of the commercial quality of Family II oils in the Central Basin

Parameter		Oil fields					
		Brujo	Catalina	Jarahueca	Jatibonico	Pina	Total
°API	Average	43.40	38.00	37.54	14.20	29.46	32.93
	Max	45.90	51.70	42.20	14.70	41.40	51.70
	Min	40.90	30.50	31.70	13.70	13.40	13.40
%S	Prom	0.47	0.66	0.76	1.68	1.87	1.09
	Max	0.53	1.80	2.03	1.75	3.89	3.89
	Min	0.41	0.09	0.20	1.61	0.40	0.09

5.5.2   Block 7

In this area, all the oil fields have low thermal evolutions (Fig. 5.6), indicating low levels of thermal maturity of the source rocks that generated them. Only in the southern part of the block there are mature oils such as Pardo 4 and Madruga 3A wells. However, these coexist with low-maturity crude oils such as wells San Lázaro 1, Basilio 2, Santa Rita 1 and Cruz Verde 116. This variation in levels of thermal maturity increases the risk of finding deep accumulations, in this area, with crude of high commercial quality.

The measured values of the geothermal gradient in this area range between 22 and 23 °C/km (Fig. 5.13), but according to the trend of this parameter, they can reach 24 °C/km in the southern part of the block. On this basis, the possible accumulations of oil in the southern part of this area are protected from biodegradation from a depth of 2300 m.

In the northern part of this area, it has been demonstrated that the sulfur-rich crude oils of Family I form acidic fluids when they react with water and dissolve the carbonate reservoirs of the lower tectonic sheet (Veloz Group + Constancia Formation), increasing the permeability of these reservoirs. The presence of Family I (Pardo 4, Basilio 2, San Lázaro 1 and Santa Rita 1) in the southern part of the block, with sulfurous crudes of up to 3% sulfur (San Lázaro 1) confirms the presence of oils with capacities for dissolving the carbonate reservoirs in the southern part of this area.

The presence of Family I in the southern part of this area indicates the presence of the lower tectonic sheet (Veloz Group + Constancia Formation) below the CVA and the ophiolites, according to the oil–source rock correlations. This fact demonstrates the presence, in the underground of this area, of the best exploratory play known in Cuba, highlighting that it has been cut by drilling in the northern part of this block. It is worth mentioning that most of the existing oil shows above this supposed deep play are small manifestations in wells, only the Bacuranao-Cruz Verde oil field is known.

5.5.3   Block 6

In this area, as in the Central Basin, high thermal evolution oils coexist belonging to different genetic families. Family I is representing by crude from wells CHD 1X, Mariel Norte 1X, Cayajabos 3, Chacón 2 and Pacheco 2, as well as La Tomasita oil seep, while Family III is representing by the Martín Mesa oil field. Special mention should be made of the fact that the oil sample studied at CHD 1X is the most mature in Cuba. However, crude oils with low thermal evolution have also been found, such as the case of Manuela Mine (Fig. 5.6). This variation in thermal maturity levels increases the risk of finding deep accumulations in this area with high commercial quality crude.

The measured values of the geothermal gradient in this area range between 20 and 21 °C/km (Fig. 5.13), but according to the trend of this parameter they can reach 22 °C/km in the southern part of the block. On that basis, the possible accumulations of oil in this area are protected from biodegradation from a depth of 2600 m.

As previously stated, Family III oils are low in sulfur and this is one of the causes of the low reserves in this family's oil fields. However, the crudes of Family I studied in this area have high sulfur contents, even the mature ones such as the cases of CHD 1X and Mariel Norte 1X with 1.14% and 5.68% respectively. These sulfur contents indicate good capacities for dissolution of the carbonate reservoirs and consequent increase in their permeability.

In this area the presence of the lower tectonic sheet (Rosario Sur type) is demonstrated by wells (CHD 1X and Cayajabos 3), so the presence of carbonates with good reservoir properties is not a risk. However, so far, in this area, only commercial accumulations related to Family III and to the upper tectonic sheet are known.

5.5.4   Block 13

In this area, only its eastern portion was analyzed (black circle in Fig. 5.12), because it is the one with the greatest potential according to a comprehensive study carried out in 2013 and which resulted in the proposal of an exploratory well in the prospect Los Ramones (Pérez et al. 2013). In here, there are a series of oil shows, among them seven asphaltite mines, which correlate mostly with Family II of Cuban crude.

The manifestations studied present different levels of thermal evolution, two of them are mature (San Gabriel and Talaren), while the rest are less evolved (Fig. 5.14). It should be noted that the mature samples are comparable with the Pina and Brujo oil fields, while the others with those of the NCOB (Fig. 5.6). This variation in thermal maturity levels increases the risk of finding deep accumulations, in this area, with high commercial quality oil.

This area has a single measured value of the geothermal gradient, 22.4°C/km (Cabreras 1). But according to the trend of this parameter, at the eastern portion of Block 13, the values range from 20 to 22 °C/km (Fig. 5.13). On that basis, the possible accumulations of oil in this area are protected from biodegradation from a depth of 2600 m.

All the oil shows studied in the eastern portion of Block 13 belong to families II or I, highlighting that seven of them show typical characteristics of the two families. This fact indicates crude oils rich in sulfur and with the capacity to dissolve carbonate reservoirs with the consequent increase in their permeability.

The fact that the oils in this area belong to families II and I also indicate the presence of the lower Rosario Sur type tectonic sheet (J_3^{km}-K_1^{ba}), according to the oil-source rock correlations. Also 23 km to the east, the Group Veloz outcrops in the Jarahueca tectonic window. Also, less than 1 km to the north, the geological formations of Camajuaní TSU appear. On the basis of these data, the presence of the main exploratory play in Cuba in the underground of this area is considered proven.

Fig. 5.14 Biomarker
relationships (Steranes)
showing the thermal
evolution of the crude from
Block 13

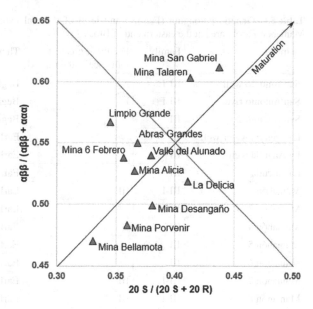

5.5.5 Blocks 17 and 18

From this area, the eastern most end of Block 17 and the western end of Block 18 (black circle in Fig. 5.12) were evaluated. The selection responds to the existence of useful data and the results of a comprehensive study carried out in 2015 (Valdivia et al. 2015). In this sector, there are many oil seeps that have in common a high degree of biodegradation. In previous studies (Delgado López et al. 2005, 2006; Pascual et al. 2008) these oil shows were classified, by saturated biomarkers, mainly in Family III of Cuban crude and of Family I in mixtures with the previous one. Later, Zulema Domínguez Sardiñas (Valdivia et al. 2015) confirmed these results from the aromatic biomarkers.

Only the samples from Holguín and Floro Pérez localities conserve the saturated biomarkers that have levels of thermal evolution similar to Cristales and Motembo oil fields, respectively (Fig. 5.6). For this reason, aromatic biomarkers were used to determine the degree of thermal evolution (Valdivia et al. 2015). In general, at the eastern end of Block 17, the samples show different thermal evolutions. Some were generated during the beginning of the generation window and the majority in a state of low thermal evolution. Only two samples, Reforma well 1 and Reforma well 2, show a high degree of thermal evolution (Table 5.6).

Valdivia et al. (2015) explained these different levels of maturity by the existence of three oil kitchen (OK):

- OK 1: It is related to the upper tectonic sheet, identified in the NCOB by wells, and similar to the Rosario Norte described by Pszczolkowski (1999). In Block

Table 5.6 Genetic belonging (Family) and level of thermal evolution of the oil seeps of the Maniabón-Farola area at the eastern end of Block 17

Samples	Family	Family of greatest influence in the mix	Thermal evolution	%S
San Antonio well	III-I	I	Beginning oil window	n.d
San Antonio river	III-I	I	Beginning oil window	4.39
San Antonio 2	III		Beginning oil window	0.77
La Anguila water well	III		Early oil window	n.d
La Anguila well I	III		Early oil window	n.d
La Anguila	III		Early oil window	0.89
Maniabón 2	III-I	III	Early oil window	2.14
Maniabón 3	III		Early oil window	1.62
Maniabón 4	III		Early oil window	1.30
Maniabón 5	III		Early oil window	1.60
Maniabón 6	III		Early oil window	0.84
Maniabón 7	III-I	III	Early oil window	3.24
Maniabón 8	III-I	III	Early oil window	4.30
Bernabé	III		Early oil window	0.74
Sta. Bárbara	III-I	I	Early oil window	n.d
Reforma well 2	III-I		Peak oil window	n.d
Reforma well 1	III-I		Late oil window	n.d

17, this sheet is associated to the Carmita Formation (K_2^{c-t}), of Placetas TSU. It generates hydrocarbons of Family III of early maturity level.

- OK 2: It is related to the lower tectonic sheet identified in the NCOB by wells, which is similar to Rosario Sur described by Pszczolkowski (1999). In Block 17, this sheet is associated to the Veloz Group (J_3^{km}-K_1^{ba}), of Placetas TSU. It generates Family I hydrocarbons during the start of the oil window.
- OK 3: It is related to the deepest tectonic sheet, Camajuaní TSU, specifically to the Jaguita (J_3^t) and Margarita (K_1^{be-ba}) formations. It generates Family I hydrocarbons with a high degree of thermal evolution.

Based on these criteria, in this area, there is a risk of finding oils with low thermal evolution in the main exploratory play in Cuba (Veloz Group). Its existence in the area is deduced by the presence of Family I of Cuban crude oil, related to the second oil kitchen postulated by Valdivia et al. (2015). The possibility of finding mature oil is related to the Camajuaní TSU, which has not shown large volumes of reserves to date.

The value of the geothermal gradient measured in this area is 18 °C/km (Farola Norte 1 well) and according to its trend the values range between 20 and 21 °C/km. These data indicate that possible oil fields would be protected from biodegradation at depths greater than 2700 m.

Table 5.6 shows that the sulfur content in the samples that have a mixture with Family I is high, demonstrating the potential for the formation of acidic fluids for dissolving carbonate reservoirs and a secondary increase of permeability.

5.6 Conclusions

1. According to the oil-source rock correlations, the presence of a specific family of oil in an area indicates the presence in depth of a specific tectonic sheet:

    - Families I and II indicate the presence of the Rosario Sur type lower tectonic sheet (J_3-K_1).
    - Family III indicates the presence of the upper tectonic sheet type Rosario Norte (K_1-K_2).

2. The production, geological and petrophysical data prove that, in the northern Cuban petroleum province, the deep basin sediments of age $J_3{}^{km}$-$K_1{}^{ba}$ type Veloz Group, in addition to source rocks, constitute the main reservoirs because they have higher secondary porosity caused by the thrusts.

3. The geological, geochemical and production data indicate that the premises that an area must meet for the existence of large accumulations of high-quality oil are:

    - Oil with a high thermal evolution resulting from a high level of thermal maturity of the source rock that generated it.
    - Oil protected from biodegradation (Temperatures > 80° C).
    - Oil must contain sulfur to guarantee the formation of acidic fluids that increase the permeability of carbonate reservoirs by dissolution.
    - Reservoir, carbonates of Upper Jurassic–Lower Cretaceous age, Veloz Group type (best play demonstrated in Cuba).

4. Of the areas evaluated (Blocks 21–23, Block 7, Block 6, Block 13 and Blocks 17–18), the first (Central Basin) is the one that meets all the premises for the presence of large accumulations of oil from high quality.

5. In the rest of the evaluated areas (Block 7, Block 6, Block 13 and Blocks 17–18), there is variation in the level of thermal maturity, which increases the risk of finding deep accumulations with high-quality commercial crude oil.

6. In Block 17, there is also the risk that the possibility of finding mature oil is related to the Camajuaní TSU, reservoirs that, to date, have not shown large volumes of reserves.

Acknowledgements The authors thank the Centro de Investigación del Petróleo and its Technical Archive for allowing the publication of non-confidential information on their research, as well as to Dr. Reinaldo Rojas Consuegra and to Dr. Evelio Linares Cala, researchers at this institution, for the exhaustive and rigorous review of the manuscript.

References

Delgado López O, Pascual O, López Rivera JG, López Quintero JO, Linares E, Sosa C (2005) Actualización de la información geoquímica y del potencial de hidrocarburos en Cuba para el año 2005. Unpublished report, Centro de Investigaciones del Petróleo, La Habana (Informe Interno), p 59

Delgado López O, Pascual O, López Rivera JG, López Quintero JO, Linares E, Sosa C (2006) Actualización de la información geoquímica y del potencial de hidrocarburos en Cuba para el año 2006. Unpublished report, Centro de Investigaciones del Petróleo, La Habana, p 171

Magoon LB and Beaumont EA (1999) Petroleum systems. In: Beaumont EA, Foster NH (eds) Exploring for oil and gas traps. American Association of Petroleum Geologist, Oklahoma, pp 3.1–3.34

Magoon LB, Dow WG (1994) Petroleum System-from source to trap. AAPG Mem 60:3–24

Pascual O, López Rivera JG, López Quintero JO, Domínguez Sardiñas Z, Delgado López O, Barreras V, Sosa C (2001) Calidad de los petróleos en los yacimientos de Cuba. Unpublished report, Centro de Investigaciones del Petróleo, La Habana, p 55

Pascual O, Delgado López O, Domínguez Sardiñas Z, López Rivera JG, López Quintero JO, Lafita C (2008) Características y Variaciones Geoquímicas en Petróleos de la Familia III. Implicaciones para la Exploración Petrolera. Unpublished report, Centro de Investigaciones del Petróleo, La Habana, p 40

Pérez Y, Valdivia CM, Delgado López O, Pérez JL, Pról JL, Cruz Toledo R, Rodríguez O, Veiga C, Perera C, Gómez J, Jiménez de la Fuente L, Domínguez Sardiñas Z, Torres M (2013). Informe final sobre fundamentación de pozo en el Bloque 13. Unpublished report, Centro de Investigaciones del Petróleo, La Habana (Informe Interno), p 83

Pszczółkowski A (1999) The exposed passive Margin of North America in western Cuba. In Mann P (ed) Sedimentary basins of the world, vol 4. Caribbean Basins, p 93

Requejo AG and Halpern HI (1990) A geochemical study of oils from the South Pass 61 Field, offshore Louisiana. In: Schumacher D, Perkins BF (eds) Gulf Coast oils and gases; their characteristics, origin, distribution, and exploration and production significance. Pennzoil Company, Houston, TX, United States

Rocha Mello M and Trindade LA (1996) Nuevas metas exploratorias en cuencas latinoamericanas de aguas profundas: Cómo proceder con el concepto de sistema petrolero. Oil Gas J Rev Latinoamericana 2(1):25–31

Sofer Z (1988) Biomarkers and carbon isotopes of oils in the Jurassic Smackover trend of the Gulf Coast states, U.S.A. Organ Geochem 12(5):421–432

Thompson Keith FM (1990) Classification of offshore Gulf of Mexico oils and gas-condensates. In: Schumacher D, Perkins BF (eds) Gulf Coast oils and gases; their characteristics, origin, distribution, and exploration and production significance. Pennzoil Company, Houston, TX, United States

Valdivia CM, Veiga C, Martínez E, Delgado López O, Domínguez Sardiñas Z, Pardo Echarte ME, Jiménez de la Fuente L, Cruz Toledo R, Gómez J, Rosell, Rodríguez O (2015) Informe de resultados de la evaluación del potencial de hidrocarburos del Bloque 17. (Unpublished report). Inédito, Centro de Investigaciones del Petróleo, La Habana, p 114

Valladares S, Álvarez J, Segura R, García R, Fernández J, Toucet S, Villavicencio B, Núñez C (1996) Atlas de Reservorios Carbonatados de Cuba. Unpublished report, Centro de Investigaciones del Petróleo, La Habana, p 247

Printed in the United States
by Baker & Taylor Publisher Services

Printed in the United States
by Baker & Taylor Publisher Services